Can a Guy Get Pregnant?

Scientific Answers
to Everyday
(and Not-So-Everyday)
Questions

Bill Sones
Rich Sones, Ph.D.
with Barb Sones

Illustrated by Dan Galas

Pi Press
New York

PI PRESS

An imprint of Pearson Education, Inc.
1185 Avenue of the Americas, New York, New York 10036

Pi Press offers discounts for bulk purchases. For more information, please contact U.S. Corporate and Government Sales, 1-800-382-3419, corpsales@pearsontechgroup.com. For sales outside the U.S., please contact International Sales at international@pearsoned.com.

Printed in the United States of America

First Printing

Library of Congress Cataloging-in-Publication Data
Sones, Bill.
Can a guy get pregnant? : scientific answers to everyday
(and not-so-everyday) questions / Bill Sones, Rich Sones.
    p. cm.
  Includes bibliographical references.
  ISBN 0-13-220695-1
  1. Science--Miscellanea. I. Sones, Rich. II. Title.
  Q173.S69 2006
  500--dc22

                                          2005020024

Pi Press books are listed at www.pipress.net.

ISBN 0-13-220695-1

Pearson Education LTD.
Pearson Education Australia PTY, Limited.
Pearson Education Singapore, Pte. Ltd.
Pearson Education North Asia, Ltd.
Pearson Education Canada, Ltd.
Pearson Educatión de Mexico, S.A. de C.V.
Pearson Education—Japan
Pearson Education Malaysia, Pte. Ltd.

CONTENTS

Part 2: *Love*   **59**

Part 3: *Death* **93**

Part 4: *Animals* 127

On a workroom wall in a Cleveland home is a four-by-six-foot map of the world with some 60 colored pins stuck in, clustered in the United States and Canada but scattered over nations of six continents— Zambia, South Africa, New Zealand, the United Arab Emirates, Sri Lanka, Ireland, Grenada, Trinidad.... These markers represent subscribing newspapers, magazines, and Web sites to our weekly Q-&-A column "Strange But True," begun in 1996 as an outgrowth of weekly beer-and-bull sessions with friends at a local pub. Our Thursday night table was reserved for "the philosophers." One evening, someone in the gang said, "You guys ought to write some of this stuff up," and "Strange But True" was born.

We had no idea it would blossom worldwide, win an invitation from The Los Angeles Times Syndicate to come on board, get picked up by *Boys' Life* magazine (kids' version), and keep email flowing from the four corners at all hours of the day and night.

It's an Internet-driven phenomenon, we've remarked many times. All the columns are sent out via email, as are questions to researchers. We garnered subscribers through painstaking spamming of long lists of newspapers and magazines. New ideas for items are often no farther away than the next site-surfing wave or emailed questions from readers.

So, whaddaya want to know? Could a voodoo curse kill you? (Possibly.) Do dogs watch TV? (Two woofs = "Yes.") Does a guillotined head feel itself hit the ground? (At least sometimes—the experiment's been done!) Do blind people see in their dreams? (If they once had sight, then yes.) Can a penny thrown from the top of the Empire State Building bore through a human skull? (No—so don't bother trying.) On a hot desert, could you drink your urine to survive? (Bottoms up!) Are there people who never pass wind? (They may think so, but…)

We have always believed that a big part of life is the Amazement/ Amusement Quotient. Our credo for the column has been that everything be documented or at least defensible as a speculation by scientific sources. So, if you're looking for suggestions that ETs are listening in on your phone conversations or dead friends are coming back for a visit, watch TV. But if you want to know about real phenomena from the real world, we invite you to read on.

The opportunity for a book came to us in a most serendipitous manner. Always on the lookout for offbeat sources for "Strange But True," we emailed Pi Press with our promotional pitch explaining our column, and we requested courtesy copies of books by biomechanics expert R. McNeill Alexander and Harvard dreams psychologist J. Allan Hobson. The email response said the books were on their way. End of story, or so we thought. Just days later, we got a call from Pi Press associate editor Jeff Galas. Based on his reading of the promo and his net surfing of SBT columns archived on various Web sites, Jeff felt that we had the makings for a popular book.

Honored and delighted, we wanted to move forward. We decided that all book Q&A's would be drawn from previously written columns. That understood, we agreed on the topic categories—four in all, covering Love, The Body, Animals, and Death—and agreed on the inclusion of illustrations for about a quarter of the items, the publication date of November 2005, and the title, *Can a Guy Get Pregnant: Scientific Answers to Everyday and Not-So-Everyday Questions*.

We had eight years of material to choose from; we had our own favorites and we also had feedback from editors, readers, and researchers. The artist's drawings shaped up fast, engaging and whimsical—just the right touch. And Jeff has proved to be as gracious and competent an editor to work with as any writers could wish for.

The book and column have been both challenging and enormously rewarding. We've fielded a delightful galaxy of questions, and we are forever indebted to readers like the one who wondered if he would have been a different person had his parents waited just five more minutes before having the sex that conceived him.

A special sort of gratitude goes out to the many, many researchers at colleges and universities all around the world, who have so generously proofed and revised our answers, helping to keep things accurate. In many cases, they took bare questions—in subject areas as diverse as anthropology, cosmology, linguistics, sports science, and zoology—and wrote wonderfully detailed answers for us. Their love for their subject came shining through, and much of the best of this book we owe to them.

Nothing has been more gratifying than to have these dedicated educators express their own gratitude to us for "helping to get valid information out to the public, countering popular mythology." One of our favorite comments came from a Harvard psychologist whom we had contacted several times over the years, "To me, you guys are gods of the weird and wonderful. Keep it up." Thank you, Dr. C.

Finally, we offer here the same invitation that we include at the end of every column: Send along a strange question to strangetrue@compuserve.com and we'll try to answer it for our weekly columns.

Bill Sones and Rich Sones, Ph.D.

THE BODY

Part 1

Q: Do humans have white and dark meat, like chickens?

A chicken's white breast muscles provide quick power for flight but tire quickly, says Ohio State University physiologist Jon Linderman. The dark leg muscles are for endurance, enabling it to walk for hours. This is reversed in ducks or geese that must migrate long distances. Fish have red, pink, and white muscles—designed for slow, intermediate, or fast action.

Any well-fed cannibal can tell you humans too have white and dark meat, though not as pronounced as fish or fowl and varying with the specimen. In a white meat mood? Look for a sprinter or power lifter, with predominantly pale fast-twitch muscle fibers. You'll find redder meat—more capillary rich and adept at oxygenation—in the red slow-twitch muscle fibers of endurance athletes such as cyclists and marathoners.

Can people grow horns?

In Herman Melville's novel *Whitejacket*, there's mention of an old woman from whose forehead grows a "hideous crumpled horn, like that of a ram." Was this just Melville's literary imagination, or could this woman really have a horn sticking out of her head?

Likely it was reality, as horned people have always existed, says Jan Bondeson, M.D. The largest horn ever recorded belonged to Mexican Paul Rodrigues, who always kept his head covered until one day at work in 1820 a falling barrel knocked him unconscious. When bystanders removed the covering, they discovered three horny growths textured like a ram's horns, some 14 inches around. He recovered, and his case was written up in a medical journal.

Modern dermatology says horns can arise anywhere on the body, consisting of concentric layers of epithelial cells. Some 65 percent are benign, caused by skin warts or other diseases. "Unlike true horns, they have no bony core."

In a famous case from the 1930s, Frenchwoman Mme. Dimanche—known as "Mother Horn"—bore a 10-incher so heavy she grew fatigued from carrying it. Many surgeons offered to remove it, but she always refused. Finally, nearing age 80, she consented because she feared meeting her Maker "with such a satanic ornament on her face."

Can people commit crimes in their sleep?

Sleepcrime is more widespread than was originally thought. Arising out of sleep disturbance or dysfunction, it involves physical harm to people or destruction of property, reports the Newsletter of the American Academy of Psychiatry and the Law.

Sleep-associated violence can occur during night terrors and sleepwalking, and is possibly related to incomplete arousal from non-dreaming sleep. Medications, alcohol, or other recreational drugs may be the cause, or psychosocial stress the trigger. And "REM-sleep-behavior disorder" might involve acting out dreams, with hallucinogenic overtones.

"Sleep drunkenness" can occur during the transition between sleep and wakefulness, when sudden arousal causes confusion and disorientation. This can result in serious aggression, even homicide.

Fortunately, most sleepcrime is not so serious. Someone might throw a sleepy punch. A guy who walked stark naked up and down a communal balcony, apparently unaware of his neighbors' presence, was convicted of indecent exposure. However, a wife who stabbed her husband 15 times—he survived—claimed to have been asleep, and the charges were dropped. Clearly, the law has a tough time knowing how to handle the shadowy cases of somnambulistic culpability.

Do blind people see in their dreams?

People born blind have no visual imagery to play back, though they dream richly in the other senses. But people blinded by accident or disease continue to have seeing dreams, which may take on special emotional significance, says State University of New York psychologist Raymond E. Rainville, himself blinded at the age of 25.

A common type is the "undoing" dream, such as one told to Rainville in which the dreamer revisited the accident that led to his blindness. In the dream, he threw himself on the car floor in time to avoid serious injury. From here, he safely watched the colliding truck shear off the car roof.

In reminiscence dreams—common during the first year of blindness—there is a happy reliving of earlier sight-rich experiences. A dreamer might watch a procession of beautifully dressed women pass by or visually feast on a huge, piping hot pizza cut into squares. "In my experience, these occur following some emotionally provocative event in waking life."

On some deep level, the dreams may help not only in adjustment to trauma but also information processing. For instance, Rainville says that when his daughter gets her hair cut short, "I will Braille it, appreciate it, comment on it. However, next time I think of her, in my spontaneous waking image, she will still have long hair. Once I dream of her in her new hairdo—that is, once I have *seen* her—she will appear to me pretty much consistently from then on."

Can dreams foretell the future?

In one sense, they obviously can: If you dream of arguing with your boss, this could reflect tensions in your relationship, foreshadowing even worse problems ahead. But don't hold out hope that you'll get rich by dreaming of next week's lottery numbers.

Mathematician John Allen Paulos told of dreaming as a kid that he hit a grand-slam home run. Two days later, he hit an actual bases-loaded triple. Coincidence? Paulos figured it was. For of the thousands of wish-fulfillment dreams over a lifetime, even if they're 1-in-10,000 longshots, chance alone will assure that a few come true.

More remarkable was a sleep-laboratory dream recounted by Dr. Peretz Lavie. On most nights, sleeper R. reported long, detailed, logical-sounding dreams, but one night he awoke to report nothing except the word "carbide," which he said "stuck in his mind." Three days later, the worst industrial accident in history occurred in Bhopal, India, killing 4,000 and injuring 20,000 more—at Union Carbide Company!

Hearing the news, Lavie was stunned. Had it happened to anyone else, he says, he wouldn't have believed it. "I have no convincing explanation, and so it joins the other reports of unusual dreams in the scientific literature, dreams which provide evidence of the multifaceted character of the abundant world created in our brains each night."

Do Siamese twins die together?

Chang and Eng Bunker, born in 1811 joined at the chest, were the original "Siamese twins." Separation wasn't possible at the time, but this didn't stop them from marrying sisters and fathering 21 children. One night in 1874, Eng awoke with a strange sensation to discover that Chang had died. Within hours, Eng too was dead, probably having bled to death as his blood pooled in his dead brother's body.

Twins Masha and Dasha of the 1960s were surgically inseparable, each with two arms but only three legs between them—two good ones and a vestigial third. Masha controlled the right leg and Dasha the left. Each had her own stomach but they shared a lower intestine and rectum. They would become ill separately and fall asleep separately.

"Conjoined" babies—1 in every 50,000 to 100,000 births, usually girls—result when an early embryo fails to split completely. Sadly, most are stillborn or die within 24 hours. Nowadays, separation is usually attempted, depending on the organs shared, though often one of the twins dies. Surgical inseparables generally die together.

"In bizarre cases, one twin will be complete while the other appears as an extra arm or head," reports Kay Cassill, director of the Twins Foundation. The 'parasite' may even be totally within the other, like a cyst or tumor ranging from a complete fetus to a tennis-ball-sized growth containing teeth and hair. "In 1972, a six-week-old child was found to be carrying its unborn sibling in its abdomen."

Is it possible to will warts away?

Eminent doctor and writer Lewis Thomas wrote about cases where people afflicted with large numbers of warts were able, under hypnosis, to "think them" off their body.

Warts have always been mysterious. Virus induced, they frequently just disappear after long stays on the carrier's body.

Adding more mystery, nine out of 14 subjects in one clinical study were able, under hypnotic suggestion, to clear up all the warts on one side of the body—as directed by the hypnotist—while leaving warts on the other side intact.

One of the subjects confused left and right and destroyed all the warts on the "wrong" side of the body.

Nobody has the slightest clue how this works, said Thomas, but at some level, matters of cell rejection are involved, including how viruses get identified as being "foreign"—all done subconsciously. There's an amazing sort of "superintelligence" within us that is far wiser than any of our current technological know-how.

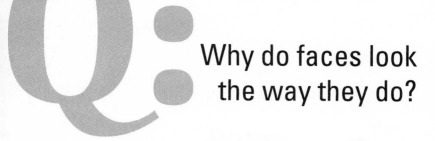

Why do faces look the way they do?

Six hundred million years ago, the human face-to-be was little more than the leading edge of a hungry tube swimming in the sea, says anthropologist David Givens.

"Picture the business end of a vacuum-cleaner hose and you'll understand the basic ancestral structure of the human face." Eventually, primitive fish faces emerged—two eyes and a snout set over a mouth. We owe our basic eyes-nose-mouth architecture to fish, but the archaic fish face was expressionless and incapable of courtship.

Little furry warm-blooded mammals took faces one step better, freeing them up and making them mobile. For the first time in evolutionary history, animals could wink, grin, and leer.

Real facial liberation came with the primates. During courtship, primate lips compress, brows lower, ears flatten, eyes widen, teeth bare, and tongues show. The zenith is the human face. "One has only to think of Rudolph Valentino's eyes, Marilyn Monroe's sugary pout, the otherworldly mouth of da Vinci's Mona Lisa, and the painful last look Bergman gives Bogart in *Casablanca*. We've come a long way from the fishes."

Can you get sick from picking your nose and eating it?

Picking your nose and eating it is actually good for you. Mucus secretions are loaded with antibodies for fighting off infections, the basic purpose of a runny nose. So pickers-and-eaters may—unconsciously, one hopes—be acting as go-between for the body and the natural defenses it produces.

Yet while the dried starches and trapped dust of a booger are generally harmless, onlookers feel queasy. So why do it? This must be a powerful private habit spilling over into public, reinforced by tactile sensitivities, play-goo textures, and salty taste.

For even deeper possible roots, watch a National Geographic-type TV program, suggests Baylor College of Medicine otorhinolaryngologist Holly H. Birdsall: "You'll see non-human primates grooming each other, picking off who-knows-what from the hair and skin, and promptly putting it in their mouths. Why should we be so different?"

Where does your body heat go after it departs?

Streaming upward off your head and shoulders—and sucking along myriad particles of you like a gaseous signature—is the excess body heat you need to dump off, or risk overheating, says Penn State professor of mechanical engineering Gary Settles, director of their Gas Dynamics Laboratory.

This airborne quintessence of you—your "human thermal plume"—rises six feet or more above your head, until you start moving along, when it becomes a trailing wake. Your plume contains skin flakes that float and settle to become 70 to 90 percent of house "dust," plus hundreds of bioeffluents ranging from moisture to carbon dioxide, to salts, and not-so-bio perfumes and colognes.

And it gets far more personal than that, carrying aloft clothing fibers, traces of drugs, explosives, and lots more. As for diseases, "signs of diabetes, gangrene, some skin disorders, tuberculosis, some cancers, and many others appear in a person's thermal plume." Your plume can be aerodynamically sampled and screened non-invasively, by sucking it into a special chamber. Even skin flakes for DNA analysis are there, raising privacy issues.

"All warm-blooded animals have plumes," says Settles. "I've seen the very impressive plume coming off a horse led back into the barn after a brisk ride. To see your own plume, just step out of a hot shower or bath into chilly air. The moisture condensation will reveal it to you."

Why do guys have nipples?

For the same reason they have brains, hearts, and nose hair. Nipples are part of the human genetic blueprint, says Cornell University comparative embryologist Drew M. Noden. Economical Nature doesn't put together two distinct sexes from scratch, but rather two versions of the same early proto-body tweaked in two different directions by sex-hormonal surges in early fetal development. Saves on parts.

Nipples kick in around the fifth week, well before the great sexual divide. Even in males, nipples and mammary glands are potentially functional, and occasionally a newborn boy under the influence of elevated maternal estrogen at birth and prolactin in Mom's milk will discharge a whitish "witches' milk" off and on for several weeks.

In a few cases, adult males subjected to high-enough doses of estrogen have actually lactated.

The nipple saga hardly ends here, adds Noden. Many mammalian species have multiple pairs of nipples running along mammary ridges on the chest and abdomen. This hidden potential remains with humans as well, and in about one baby in a hundred (more commonly in boys), an extra "mole" or so will show up. In rare instances, females at puberty end up with several developed breasts—the record being 10, all milk-producing.

Is it possible to stutter in one language but speak normally in another?

Yes, says University of Pittsburgh stuttering clinician J. Scott Yaruss. It can run both ways—there are stutterers who find they stutter less or not at all when learning a new language. When their familiarity with the new language grows, on comes the stuttering. Others stutter more at first, which then tapers off as familiarity grows.

Stuttering can be strange, echoes Auburn University fluency disorders specialist Larry Molt. It reflects the deep complexity of speech itself. Witness those folks who stutter in their everyday talk but who become fluent when they break into song. Or certain well-known actors who conquer their stuttering best when they adopt a role and "speak in character."

"Nobody understands this well, but neuro-imaging research seems to indicate stuttering flows out of different brain circuitry than normal speech," adds Molt. "And since later-learned languages are handled differently by the brain, either acquisition or storage, this may be how some bilinguals keep at least one tongue stutter-free."

Can a person's hair turn all white overnight?

Marie Antoinette's hair was said to have turned all white overnight from fright, before she was executed during the French Revolution. The same has been said of Thomas More, who was executed in England during the reign of Henry VIII. Overnight graying or whitening has been reported for centuries, but the two historic cases cited are disputed.

What does modern science say? Will a strong scare really de-color the hair? No and yes. The "no" is that hairs don't suddenly just lose their pigment. The "yes" is that a strong emotional shock can cause a rapid loss of large amounts of hair, and if this is mostly the pepper of a salt-and-pepper head, then salt shall remain.

Losing mostly pepper (still pigment-bearing locks) is indeed possible with a condition termed "alopecia areata." So for these folks, a life crisis can suddenly leave them with a whiter shock of hair.

Q: Are there people who don't have "flatulent moments"?

A: We heard a guy bragging about the fact that his wife had never "passed wind." This didn't seem right, so we consulted the experts....

"I don't believe this is possible," said Australian gastroenterologist Terry Bolin. He had 60 men and 60 women record their emissions over a three-day period, and three per day was the low, 38 per day the high. "The peculiar lady in question may pass gas unconsciously during sleep."

Other G.I. docs were less diplomatic: "Has she had her sinuses checked? How about her hearing?"

As for her husband's assertion, UCLA professor of medicine David Diehl said, "I can only say that he can't monitor her flatus 24 hours a day! Despite how good sphincter control one might have, the gas must eventually pass."

Showing just how good such control can get, the Frenchman Joseph Pujol became a popular Moulin Rouge act in the 1890s by mastering withholding his gas, then releasing it on cue to imitate the sounds of creaking doors, amorous bullfrogs, hooting owls, ripping cloth, machine guns, cannons, and more. The elite doctors of the day examined Pujol, all amazed he could sit in a tub of water, draw in several liters, then expel a liquid jet that traveled five meters across the room.

Can dreams help you stop smoking or lose weight?

Strange as it sounds, they just might. When people in a stop-smoking or stop-drinking program were asked to describe the dreams they were having, a third of them reported DAMIT dreams—"dreams of absent-minded transgression," says psychologist Robert A. Baron of Rensselaer Polytechnic Institute.

Here the dreamers suddenly became aware they had reverted to their old bad habits without wanting to, threatening all their good efforts. Many awoke in a panic, relieved it was only a dream.

And the most surprising finding: The subjects who reported having such nightmares, in effect visualizing the emotional cost of failing, were the ones most likely to wind up successfully kicking their addictions.

Is DIY ("Do It Yourself") surgery possible?

Necessity is the mother of much, including do-it-yourself surgery. Most of the self-surgery literature involves amputations, says Columbia University surgeon Allan Stewart. In 2003, a 27-year-old mountaineer amputated his own arm with a pocketknife after being trapped under a boulder for five days in a remote canyon in eastern Utah. Self-amputation is not unique to humans; coyotes caught in a trap will bite off an extremity to avoid capture.

In a rare case, a 40-year-old woman in Mexico performed a crude C-section on herself when she went into labor and ran into trouble. She had previously lost a baby in childbirth and feared this might happen again. After three glasses of hard liquor, she cut through skin, fat and muscle, reached inside, and pulled the baby out. She cut the umbilical cord, then passed out. Amazingly, both Mom and baby survived.

In a very different case, a surgeon at the South Pole out of necessity did her own biopsy, made a fateful diagnosis, and wrote a book about it.

And here's San Francisco surgeon Carlos Corvera on self-surgery: "I have in fact sutured a laceration on my hand. Of course I used local anesthetic. Once you get over the issue of cutting into flesh, I doubt it matters much that it is your own."

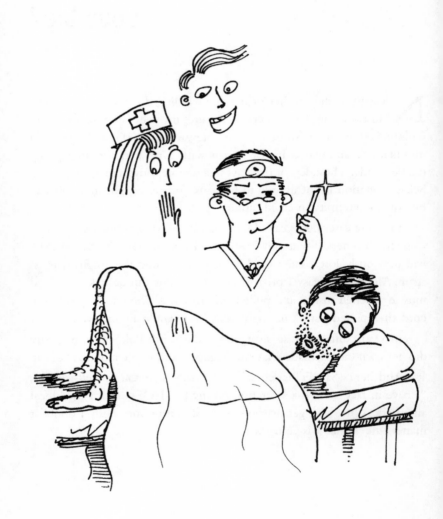

Can a guy get pregnant?

Well, amazingly enough, yes, in the sense of carrying to term an already fertilized egg implanted in the body. David Bodanis, science writer and former lecturer at Oxford University, has written about how the egg would need some sort of warm moist tissue to latch onto, with a suitable blood supply. In women, there have been cases where a fertilized egg slipped out of the fallopian tube, floated around in the abdominal cavity, and attached to the pelvic wall. Nine months later, a healthy baby was delivered by cesarean section.

"There's no good reason why a properly implanted and hormone-supplied zygote could not enjoy the same normal nine-month growth inside a man," he says.

Ronald R. Magness, director of perinatal research at the University of Wisconsin-Madison, explains that women shortly after conception don't need their ovaries for progesterone because it is produced by the human placenta. This hormone functions largely to keep uterine muscles relaxed, something a guy wouldn't have to worry about, not having a uterus.

"Implantation of the highly invasive conceptus would likely occur in the greater omental fat pad, the bladder, the peritoneum, or maybe on or near the kidney capsule. With exogenous progesterone treatment, the likelihood of this working should be greatly improved."

Knew those fat pads could turn out to be useful.

Is there more human flesh on Earth than any other kind?

O f all large animals on the Earth today, humans account for the greatest total biomass. Add up the body bulk of the Earth's 6 billion human beings and you're talking maybe 400 million people-tons. To transport all of us at once, you'd need 40,000 10,000-ton ships.

Probably the closest competitor is our old barnyard friend, the cow. But at only about a billion globally, it's doubtful that they constitute a beefier bunch.

How sensitive are the five senses?

A s a typical healthy perceiver, you are able to...

- Smell a lone drop of perfume diffused through a three-room apartment, says University of Michigan psychology professor Charles G. Morris.

- Identify something as salty or sweet within .1 second of it touching your tongue.

- Gauge a sound's direction based on a .00003 second difference in arrival time from ear to ear.

- Feel the weight of a bee's wing falling on your cheek from the height of .4 inch.

- See a small candle flame from 30 miles away on a clear, dark night.

- Taste .04 ounce of table salt dissolved in 530 quarts of water (employing a fraction of your 10,000 taste buds).

- Feel a nerve fire when a downy skin hair vibrates .00004 inch.

- Hear frequencies from 16 to 20,000 cycles per second (with middle C at 256)—a range of more than 10 octaves—before your hearing begins to deteriorate with age.

What's the trick to ear wiggling?

It's in the genes and runs in families, like the ability to curl one's tongue, says ear wiggler R. Steven Ackley, communication disorders specialist at Gallaudet University.

The three extrinsic muscles of the ear are probably becoming vestigial in humans. Moving the ears and tightening the forehead could have been part of how early man bluffed challengers for a mate, much like other primates do. To be sure, we do not need to orient our pinnas toward a sound source the way wolves do. Some of us (throwbacks) can still contract at least the "auricularis posterior," which pulls the pinnas back. "But give our species another hundred thousand years or so and there may be no more Stan Laurels among us."

Why do you see bright lights when you press on your closed eyelids?

Pressing on your closed eyelids directly stimulates the optic nerve, causing these "inner" sight sensations.

Lack of stimulation will also trigger them. People confined to dark cells report seeing phosphenes ("prisoners' cinema"), as do truck drivers after staring for long periods at snow-covered roads, says Cleveland State University's Jearl Walker.

Back in Ben Franklin's day, it was fashionable to hold phosphene parties (Franklin himself attended one) at which guests would join hands in a large circle, and then receive a high-voltage shock from an electrostatic charge generator. Switching currents on and off unleashed colorful phosphenes in the participants (don't try this!).

In experiments by Paul Tobias and J. P. Meehan, reported in *Scientific American*, blindfolded volunteers who were spun in a centrifuge saw phosphene arrays of blue spots and stars. At 3.6 G's (3.6 times gravity), these inner visions became "golden worms," then at 4.5 G's the worms transformed into brilliant orange geometric patterns that began to pulsate. Following the centrifuge ride, subjects described phosphene afterimages lasting a minute or more, weird donut-like shapes, or appearances of a solar eclipse against a dark background.

Given the acceleration-phosphene connection, pilots and astronauts are also apt to experience these images, as are rapidly spinning amusement park riders—providing, adds Walker, they're not too white-knuckled to notice the show.

Are sleepwalkers acting out their dreams?

Actually, it's quite the opposite. During dreaming sleep, people are actually paralyzed, capable of jerks and twitches but little body movement. Without this motor shutdown function, the nocturnal world might be inhabited by peripatetic dreamers fleeing imagined monsters or trying to fly. So, when someone is sleepwalking, you know he or she isn't dreaming but is rather in a deeper stage of sleep (stage 3 or stage 4).

Somnambulists will generally sit up, get out of bed, and go for a stroll, says pioneering sleep researcher Peretz Lavie, M.D. Their eyes are wide open but unseeing; they find their way by memory and thus may trip over unexpected obstacles. Sleepwalking is a form of "automatism," where learned behaviors are repeated.

In one case that puzzled Lavie, a six-year-old boy would get up and go as far as the lab's attached electrodes would allow. "Then he began to wave his hands around with strange and exotic movements." When his parents saw the video, they revealed that this was just part of a school play where the boy played the role of the sun.

Will it ever be possible to freeze people and bring them back to life in the future?

Cryonic suspension, or "solid-state hypothermia," involves freezing a fresh corpse in hopes of reanimating the person at a later time, says Kenneth V. Iserson, M.D.

In case you haven't explored the options with your local cryonics society, you can have whole-body suspension starting at $120,000, or cut-rate neurosuspension where only the head is frozen for around $50,000 (in which case your caretakers will need to find a suitable body for you later, or a mechanical contrivance for your hookup as a cyborg). To date, there are around 100 "souls on ice," plus a few dozen pets.

The freezing is the easy part, though you need an intact cadaver and high-tech equipment at the ready within half an hour after death. A temperature of around –320 degrees F must be maintained using liquid nitrogen.

Nobody has any idea when we'll be able to bring frozen bodies back to life. At present, there isn't even the faintest glimmer of scientific know-how for this, emphasizes Iserson. Moreover, the freezing process itself does radical damage to the cells. The same reason why frozen vegetables don't have the same texture as fresh ones applies here—freezing water expands and blasts out the cell walls.

What you are preserving, remarked one critic, is not living tissue but meat. "To expect to bring it back to life is like believing you can remake a cow out of hamburger."

Could a modern human beat a Neanderthal in a wrestling match?

Let's assume that both the modern human and the Neanderthal are both young and healthy. "I'd put my money and as much of anyone else's as I could find on the Neanderthal. They were incredibly strong in their upper bodies and heavily built," says Washington State University anthropologist Tim Kohler.

But if you changed the contest from a one-on-one to a situation calling for cooperation and teamwork rather than brute strength, then the betting would have to go with modern populations. That's probably the factor where moderns had a slight advantage over the Neanderthals—why we're here today to wrestle with life and they're not.

Can you lift a person by the hair without the hair ripping out?

"I've seen Chinese circus performers do this," says Kevin J. McElwee of Philipp University, Dermatology Department, Marburg, Germany. Healthy hair is strong, about equivalent to copper wire of the same diameter.

So the question is how well the hairs are anchored in the scalp skin—something that hasn't been studied in depth. But chest hairs have been tested. They hold roughly 70 grams each, especially under a slow "pluck," akin to the hanging feat.

Assuming head hair to be similarly rooted, and multiplying 70 grams by 80,000 to 120,000 scalp hairs per person, the math works out to 5,600 to 8,400 kilograms hanging force per scalp—that's 12,000 to 20,000 pounds!

But here's a warning for anyone about to try to hang a car from their hair: "In reality, the max would be less, due to imperfect fibers, weathering defects, etc., but still plenty of anchoring strength overall to lift a person by the hair—and several others besides," says McElwee.

Why are sumo wrestlers so fat?

Try this, suggests author Mina Hall: Stand with legs together and have a friend try to push you over. Next, crouch down with your legs apart and do the same. Get it? The low crouch and big belly lower a wrestler's center of gravity, adding stability. And stability is all important in a sport where losing one's balance and going down means defeat.

Imagine being 200 pounds and trying to push over someone weighing 600. It can be done, but this is rare. Rookies may start at 200 pounds, but gaining weight is usually required for advancement. Top division sumotori will go at 335 and up.

But big doesn't mean fat. A layer of fat does help cushion blows and falls, "but underneath that soft tissue lies rock-hard muscles... Former yokozuna Chiyonofuji had only 11 percent body fat—the average is 13 or more," says Hall.

Sumo big entails shopping for XXXL clothes, driving roomy vehicles (vans are popular), eating out of special large cups and bowls, and sitting on 20-by-23-inch toilet seats, half a foot larger than standard. Everybody just prays the wrestlers' "diapers" don't come off during a match, which, reasonably, is an automatic loss.

Can men breastfeed babies?

Men can "suckle" their babies with a special bottle backpack and tubing hookup running down to a "nipple" over the chest to mimic the positions used in breastfeeding. That's psychology.

On the farther fringes of biology, males could in theory produce actual milk since they have breasts containing rudimentary ducts, "but I am personally unaware if this has really been achieved or to what extent," says University of Michigan reproductive endocrinologist John Randolph, Jr., M.D. Female hormones would be required for breast development, along with intensive breast stimulation.

It is possible for men who have their testes removed or medically shut down to get normal-appearing breasts by use of estrogenic hormones, adds University of Missouri reproductive endocrinologist Steven Young, M.D., Ph.D. Transsexuals (men to women) on estrogen get an increase in fatty tissue as well as glands. Of course, the size alone is no indication of function, and "I am not aware of any reports in the English language literature of normal lactation in these men. On the other hand, I suspect that it might be possible," says Young.

Can a body live without a head?

Sure. Thank the breathtaking breakthroughs of genetic engineering. The cloning of Dolly the sheep and the creation of a headless frog embryo prompted Dr. Patrick Dixon, physician and world-renowned futurist, to forecast cloned colonies of headless humans kept as spare parts factories in the not-distant future.

Given its technical feasibility, there will be enormous economic pressure for this to happen, Dixon told the British Press Association. It will likely occur in countries where there is little or no gene legislation.

Besides, the world already has "Miracle Mike," a headless chicken that lived in the 1940s. Earmarked for supper, Mike met a bizarre fate when a poorly aimed hatchet took off his head but left neck and brain stem intact, enough to keep his body strutting about for a couple of years. Mike was fed via eyedropper through the open hole of his esophagus. *Life* magazine ran a photo of him amidst his barnyard brethren, captioned, "Chickens do not avoid Mike who, however, has shown no tendency to mate."

If you're starving, can you eat your clothes?

A traveler stranded in the unforgiving Australian outback survived by eating his leather clothing and drinking water from his car radiator, says nutritionist Clive Barnett.

If you're lucky enough to be marooned on an ice floe instead of a desert, you'll have an endless supply of water. If you're even luckier and have a knife and cooking pot handy, you can melt the ice over your fire and boil strips of your leather shoes, belt, or coat.

Another option would be to barbecue them over the flame. Leather is animal skin treated with tannic acid, and heating will improve its digestibility and tenderness.

"Of course, if you're wearing edible underwear, your problem is solved!"

Are there "false pregnancies" in men?

Abdominal swelling and discomfort, morning nausea, unusual food cravings, and breast enlargement—all were symptoms reported by patient "George" to Deirdre Barrett of Harvard Medical School.

Barrett knew of false pregnancies in women, who stop menstruating and show all the outward signs. Some husbands of pregnant women experience symptoms, such as strong stomach cramps during the wife's labor, but these men don't think they're pregnant.

George was different. Full-blown false male pregnancies are rare, and his seemed to have begun after a stop-smoking session with a hypnotist. "Picture the person you would like to be," the therapist had directed. A pregnant woman popped into George's mind. He got the idea that a male could carry a fertilized egg from a tabloid, and he latched onto the hope of somehow being a hermaphrodite with hidden female organs.

George's gay lover had recently died, and now George imagined he was carrying the man's baby. But he was pragmatic enough to accept the outcome of the medical tests: no actual pregnancy.

At this, George's abdomen began to shrink, and a few weeks later he announced to Barrett: "I'm not pregnant anymore." His new guiding imagery was of a woman after childbirth, and this he said was helping him work through missing his lover. "But I kind of wish I had kept the breasts."

Can a human glow like a firefly by eating the right food?

Fireflies emit light via a chemical reaction between luciferin and luciferase—a protein and an enzyme, says Stanford University's Michael Bachmann, M.D., D.Sc.

Genes to produce these chemicals are in a number of "bioluminescent" species, ranging from bacteria to worms and insects. Mammals, including humans, do not have such a set of genes, and so are unable to emit light naturally. "Hence, we would not glow, even if we ate pounds of the right stuff, i.e., luciferin—which is extremely expensive at $200/gram."

However, modern biotechnologists now know how to introduce into laboratory mice the firefly luciferase gene, which acts along with injected luciferin to light up target cells for disease tracking.

Some brave new "artists" apparently have plans to take this a step further to create glowing dogs and cats, either art for art's sake or for pure entertainment. "But I think it is morally wrong to subject animals to such suffering when there is no benefit to human beings or other animals."

Are Siamese triplets possible?

Conjoined twins are rare enough—1 in every 50,000 to 100,000 births.

More rare are conjoined twins born as triplets with a non-conjoined sibling, though there are a number of such citations in the Medline database, says Dr. Peter Klein of the University of Pennsylvania Medical School. Dr. Daniel Kessler, also of the UPenn Medical School, adds: "Research into the embryology of animals indicates that Siamese triplets could form, but the frequency of Siamese triplets in humans is predicted to be far lower than Siamese twins, and the live birth of Siamese triplets would be even less likely."

So it is not surprising that clear and documented cases are hard to come by, though the 19th-century Gould and Pyle's *Curiosities of Medicine* cites an 1834 case in Sicily of three boys born with a single torso, two hearts, two stomachs, two lungs, and three heads.

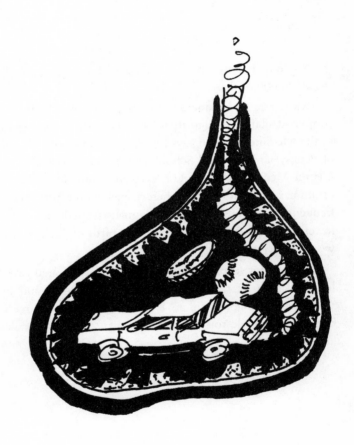

Can human stomach acid dissolve an automobile?

The hydrochloric acid of the human digestive process is so strong it will easily eat through a cotton handkerchief and even through the iron of a car body, said Isaac Asimov. The stomach walls themselves are protected by a thin film of sticky mucus.

A penny swallowed by a two-year-old was found riddled with holes four days later, says *Daybreak* magazine of the University of California-San Francisco. An ulcer had developed where the coin was lodged.

Most swallowed coins pass harmlessly, but in some cases the metals react with the acid to form toxic substances that cause inflammation. Curiously, U.S. pennies minted after 1982—of mostly zinc—have been found to cause more stomach damage than earlier ones, which are 95 percent copper, and only 5 percent zinc. Coins made of nickel do little harm in this way.

Why are there equal numbers of men and women?

Because one guy could easily impregnate many women, it seems like it's a waste of biological resources to have so many guys, doesn't it?

This is a tricky one, said Stephen Jay Gould. Imagine a society where suddenly an excess of females is born, tipping the sex ratio. Since males are now rarer, their opportunity for mating increases. Each male has the opportunity to impregnate multiple females.

Some couples tend to produce more males and some more females. At this point, there would be a selective advantage in favor of parents who produce more boys, because males have greater reproductive opportunity in this predominantly female society.

Now, as more boys are born, their numbers will begin to push the sex ratio back toward 50-50 again. The same argument in reverse would apply in any society where suddenly an excess of males was born, said Gould.

This is why it's generally a boy-meets-girl world, rather than a boy-meets-girl-&-girl-&-girl-&-girl world.

How is it possible to walk over red-hot coals in bare feet?

Fear and physics protect the feet, says physicist Jearl Walker of Cleveland State University. He should know: He has shed footwear a number of times to demonstrate firewalking to his students.

The theory: Red-hot wood coals, though at a high temperature, contain little energy. So if you quick-step across the bed, little heat gets conducted to your feet. It's like touching a cake right out of the oven. Even at 350 degrees, the cake is a poor heat conductor, so you won't get burned. But don't try touching the metal pan!

A second factor is foot moisture. Some firewalkers step through wet grass or pour water on their feet beforehand. Then, as the water heats up and evaporates, it absorbs energy from the coals. It's similar to wetting your fingers to snuff a candle.

For his own firewalks, Walker eschewed foot dousing, figuring his fear of injury would make his feet sweat enough to protect him. For his first four walks, everything went smoothly. Then, on the fifth, he got complacent, his fear left him, and his "no sweat" effort got him two very badly burned feet.

Can a head live without a body?

Yes. This is already possible with current medicine, says University of Pittsburgh neurosurgeon Hae Dong Jho, M.D., Ph.D.

It's actually been done—sort of. Decades ago, a Russian brain surgeon took a head from one monkey and attached it as a second head to another monkey's neck. The extra head functioned as an independent individual living as a parasite. "If a human head were kept alive with a machine, the head would be much like the monkey's extra head."

The first prerequisite would be intact veins and arteries for recycling blood, which must be kept well-oxygenated and laced with glucose and a few hundred other essentials, says State University of New York neurosurgeon James Holsapple. There'd be no need to breathe. The blood can be taken from the head and oxygenated, "then returned without the silliness of bellows and air moving through the head's nose, etc. Also, no need for eating!"

Though medically feasible, says Holsapple, "I feel bad for the patient." What happens to all the old neurological and hormonal feedback connections that gave the sense of having a body? What about talking, movement, sex?

"Better to just lop the head off, dunk it in quick-freeze liquid nitrogen, and wait for scientists in 5,000 years to find a way to suck the personality 'out' and stick it into something more durable—like a clone or robot brain."

Q: Can sex cure a headache?

Nice try! Although one study does mention "a rise in the endorphin rate accompanying orgasm," most research points the other way, says Dr. Erick Janssen of the Kinsey Institute. Rats and hamsters show, if anything, decreased sexual behavior under the influence of endorphins, and in humans, sex can actually trigger headaches.

There are different types of headaches—migraines, tension-types, clusters, chronic dailies, etc.—and these can't be expected to respond uniformly to sex. Arousal might diminish attention to pain, but the physical activity itself could bring on an "exertional" headache.

There is even a clinical subtype called "benign vascular sexual headache," with possibly "acute headache and hypertension occurring at the height of sexual excitement."

Can you drink your own urine if you are dying of thirst?

"**A**bsolutely. There are many such stories," says University of Virginia urologist Terry T. Turner. Some people even advocate drinking urine as therapy.

In addition to being 95 percent water, urine contains a smorgasbord of dissolved salts and other solids scientists are still cataloguing—maybe 1,000 altogether, of which only 200 are known, notes the Argonne (Illinois) National Laboratory.

Among these are vitamins, amino acids, antibodies, enzymes, hormones, antigens, proteins, immunoglobulins, uric acid, urea, proteases, and growth inhibitory factors in human cancers. "Believe it or not, scientists have known for years that urine is antibacterial, anti-protozoal, anti-fungal, anti-viral, and anti-tuberculostatic!"

LOVE

Part 2

Why is kissing so popular?

Every kiss from infancy onward reverberates with echoes of pleasure and attachment, says Leonore Tiefer of the Albert Einstein College of Medicine. "The lips and tongue have large representation in the brain—every infant must suckle to survive. As we suckle, we feel, and we don't forget."

Even in cultures where tongue kissing is disapproved, the need for security and attachment produce the eros of cheek-to-cheek contact, nibbling lips, and inhaling the aroma of the beloved's face.

With their overriding intimacy, kisses deeply bond—"It's you and me against the world." A theme of Western literature is that where people cannot choose their own mates, or free expression of sexuality is denied, kisses come to symbolize social chaos. "For Romeo and Juliet, kissing was dangerous, mortally bonding the *wrong* pair."

The kiss supreme is the wedding kiss—at once symbolic and sexy. "It's both clean and dirty, both forward looking and backward looking, both universal and particular. I love theory," says Tiefer, "but the real kisses are best."

How old do you have to be to fall in love?

Falling in love can occur as early as the capacity to step outside of one's own sense of self and attach to another person or image of that person, says California State University-Los Angeles psychologist Stuart Fischoff. Probably a first-grader, around age six, can love someone, so long as the child has moved away from being totally ego-centric. But there is no age limit on being ego-centric. So, people who are adults may be unable to truly fall in love because they can't get out of their own need system—seeing a love object as something that can give to them but not something that they want to give to.

Six-year-old love may look like puppy love, but it's the real deal, Fischoff stresses. Many people confuse lust with love and can't imagine that a six- or ten-year-old can love. "They can, even if it doesn't last a long time. Love at any age can arrive and depart in short order because it's not love that makes long-term relationships, it's liking."

Q: Do opposites attract?

People almost always wind up marrying someone who lives within a short car ride of their home, is of the same race (99 percent), same religion (75 percent), and similar degree of attractiveness (unless one of the partners "trades" extreme good looks for money), says David M. Buss, psychology professor at the University of Texas, Austin.

Nonsmokers usually pair up with nonsmokers. Outgoing people usually marry outgoing people. Argumentative people usually marry argumentative people. Couples also tend to match up based on age (within three to four years), height, politics, health, and socioeconomic status. Turns out, Buss says, that the old saying about opposites attracting is largely myth.

"Homogamy" is the word for this. Because children are often like their parents, this is just another way of saying most of us marry a guy or gal just like the one who married Mom or Dad.

Why do the "girls get prettier at closing time"?

Country music star Mickey Gilley had a hit in 1975 with his song "Don't All the Girls Get Prettier at Closing Time?" Is that true? Do the girls really get prettier at closing time?

First of all, that's only half the story: The boys get handsomer too, say psychology professors Robert Baron and Donn Byrne. It works out a lot better this way. It would hardly suit humankind's reproductive future if the gals metamorphosed into princesses while the guys stayed pumpkins.

In one study at a college bar, looks ratings dogged along at 9:00 P.M., perked up a bit by 10:30, then soared around midnight. This was not due to alcohol "goggles" either, because same-sex looks ratings remained the same. More likely it was the scarcity effect, as bar patrons realized time and opportunities were dwindling.

Does it help to practice kissing on an inanimate object?

Don't worry—you can get osculatory experience by puckering up to a pillow or poster, or by "mentally rehearsing" with a fantasy lover, says Olympic sports psychologist Dr. Jeff Simons. After all, kissing is a motor skill, and top athletes have long been perfecting their motor skills via intense imagery.

Think of someone you'd really like to kiss, and close your eyes. Try to feel it. The more senses you bring into play, the better. Because the subconscious mind cannot distinguish reality from fantasy, at some level these kisses will really be happening to you!

A word of caution: You need feedback. Are your kisses too hard, too wet, too long, too probing? You may imagine your kiss as wonderful, but would a real partner agree?

The second pitfall is unrealistic expectations. "It would be a shame to live a fantasy life to such a degree that reality is nothing but a letdown. So enjoy those imaginary kisses but don't fall in love with your fantasies. And remember to dry off the pillow before lying down to sleep."

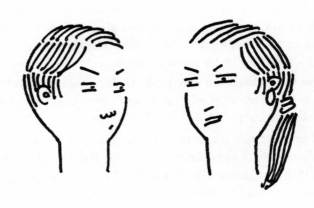

How choosy are most people when it comes to dating?

Without really thinking about it, we follow our culture's dictates: In effect, we run a quick "criteria check" on every potential partner who happens our way, say psychologists Elaine Hatfield and G. William Walster.

For most of us—if we're honest—this means ruling out anyone not close to our own age, of a different race, not sharing our educational level, etc. "The moment we get down to specifics, it becomes clear that the 'little' we ask is startlingly extensive. We want a good-looking person, a person with plenty of time to devote to us, a successful person, a sensitive person, a person with intelligence and a good sense of humor—the list goes on and on. Perfection is what we really want."

Try an experiment: Keep a tally of the number of men or women you encounter—at the office, on the street, at the supermarket—and how many of them you would be willing to date. Be honest.

"You'll be startled to find out just how incredibly fussy you really are."

Q: Does danger contribute to romance?

R esearchers had an attractive woman approach male strangers coming off a long, narrow, swaying footbridge high over a deep gorge. The women asked the men to fill out a questionnaire—supposedly for a sociology class. The men were invited to "call later for the results, if you're interested." It turned out that these interviewees were far more likely to make a call than a control group of men who met the woman right after crossing a low, safe bridge.

Upshot: Passion can ride piggy-back on fear, anxiety, or other emotional states, so take your date to see a thriller movie or for a heart-pounding roller coaster ride.

How can you tell if a marriage will last?

Some time ago a love poem in a bottle washed ashore on a Guam beach: "If, by the time this letter reaches you, I am old and gray, I know our love will be fresh as it is today.... If this note never reaches you, it will still be written in my heart that I will go to extreme means to prove my love for you. Your husband, Bob."

When the woman addressed was reached by phone and read the note, she burst out laughing. "We're divorced."

Now, how can you tell if your marriage will be one of the 50 percent in the United States that ends in divorce? Score yourself and your spouse on the following checklist: (1) Married after age 20; (2) Both from stable two-parent homes; (3) Dated a good while before marrying; (4) Both well educated; (5) Stable income from a good job; (6) Live in a small town or on a farm; (7) Did not cohabit or become pregnant before marriage; (8) Are religious; (9) Are of similar age, faith, and education.

If you answer "yes" to all these, says psychologist David Myers, you're very likely to stay together. "But if none of these is true for someone, marital breakdown is an almost sure bet."

Does height matter when it comes to love?

So strong is society's male-taller norm that in one sampling of 720 married couples, researchers John Gillis and Walter Avis found only one couple flouting the rule. Men average four inches taller, but still, assuming random matches, the sample should have included about 20 woman-taller couples.

One theory is that if men and women paired off at roughly equal heights, this would leave many very short women and very tall men unmatched. By having males stand taller than their mates by about the average height difference between the sexes, society ensures fewer will get left out biologically.

Another theory points to the male desire for dominance: "Taller is bigger, better, more powerful." This ancient sentiment, evident in business and politics, may rear its head in romance as well.

Do people flirt the same way in all cultures?

Flirting techniques are instinctive. No practice is necessary, says bestselling author Diane Ackerman. It doesn't matter if you're in Holland, Taiwan, Indonesia, or Amazonia.

"The eyes—a little glance, glance down, glance back with a little smile—that will work anywhere," says University of Michigan psychologist Phoebe Ellsworth. But a word of caution: American guys are a lot more blatant than men in many other cultures, and while the women may appreciate this, their fathers and brothers won't. You could find yourself in deep trouble by being too friendly. "In many cultures, flirting with strangers just isn't done."

Physical closeness, parallel postural orientations, and touch too are universal, but there are differences in how these are used, says University of Connecticut professor of communication sciences and psychology Ross Buck. In a famous example, couples at cafés were observed to touch 180 times per hour in San Juan, couples in Paris 110 times, and in London 0!

Why do so many couples get divorced?

L et's do the math. Suppose 80 percent of spouses are F's—"faithfuls" who try their best to keep a marriage going. The other 20 percent are N's—"nasties," impossible to get along with over the long haul. You can estimate your own percentages, but everybody knows the types.

Now, out of every 200 people marrying for the first time—100 couples—64 of the marriages, on average, will involve two F's (.8 times .8 equals .64) and will endure. An average of 32 of the 100 marriages will involve an F and an N—and eventually break up. Four of the 100 marriages will be between two N's—ill-fated from the start. The overall divorce rate for this first round is 36 marriages out of the 100, or 36 percent.

Now, Round 2: All of the F's who married F's are still together, and not in the re-marriage pool. From the 32 divorced F-and-N couples, there will now be 32 F's and 32 N's making the dating rounds. From the 4 divorced N-and-N couples, there will be 8 N's. Total: 32 F's, 40 N's. These 72 former spouses will form 36 new marriages. But only about seven of these, on average, will be F-and-F. The remaining 29 will be N-and-F, or N-and-N, and soon end. Divorce rate for Round 2? A whopping 80 percent. Overall divorce rate for both rounds will be close to 50 percent.

For Round 3, you can see the faithful F's are scarcer still, the nasty N's are in abundance, and marriage licenses might as well be printed in disappearing ink.

Can a dream lead to true love?

Love is so fickle that a lot less than a dream has made it all work. Critic and writer Alfred Alvarez tells of a divorced man in London who grimly played the field. He was kind of "interested" in an American girl. They lived together awhile, fought too much; it was on and off.

"After about two years of this, I had a dream." He was seeing another woman now, but it too was rocky, though he imagined he was in love. In the dream, he was dancing with the American girl, just like old times. "I pushed her out at arm's length and looked at her." Her hair was now white and, he realized, so was his. "We're old, I thought in my dream, and we're together—and perfectly happy."

He awoke happy and couldn't understand it. They had always enjoyed each other's company, but somehow that hadn't seemed enough, "too natural, not sufficiently doomed.... But the dream was telling me what I refused to know: that I wanted to spend the rest of my life with her."

A few days later, they ran into each other on the street and he told her of the dream. Soon they got back together, and six months later were wed. All of this Alvarez recounted 35 years later in his book, *Night*, and the couple was still married.

Just like the countless other "dream" lovers out there.

Why is love so fickle?

You can take this all the way back to the child in us, say psychologists Elaine Hatfield and G. William Walster. Psychoanalyst John Bowlby describes how a 2-year-old will check that Mom is nearby before he sallies forth, having lost interest in her. But should she disappear for a moment, it's a different story—panic and agitation. Of course, once she returns, he's off again.

Sound familiar? When we meet someone who can buffer us from life's frustrations and tribulations, we are relieved, and in love. But once we become really secure, we stop focusing on what we have—security— and start longing for what we don't have—excitement.

"Obviously, the child remains in all of us. Often it's when we're most secure in our loving relationships that we find our minds wandering to castles—and lovers—in Spain."

Q: Does music contribute to romance?

" If music be the food of love, play on," wrote Shakespeare in *Twelfth Night*.

The Bard knew his brain-body basics, says Los Angeles ethnomusicologist Elizabeth Miles. The first place music hits after coming through your ears is the hypothalamus, home of basic drives from hunger to lust. "So if a song has ever made you look at a member of the opposite sex with a fresh eye, that's why."

Next, the electrical impulses of music move through the entire nervous system, either speeding up or slowing down its function. Choose something loud and upbeat and it's a sort of musical Viagra, increasing circulation and breathing rate. Go low and slow and it's more like drinking a glass of wine. Either route can provide a libidinal boost, depending upon your disposition.

In a groundbreaking experiment by psychologist Avram Goldstein, people rated the thrills they get from music even higher than sex. But "in my own nationwide survey of how people use music in their lives, I was surprised to note a lack of effort to exploit music's erotic potential, suggesting we would all be wise to remember Shakespeare's words...and play on."

Is it better to dump your lover, or to get dumped?

It matters a lot whether you happen to the "breaker" or "breakee," say psychology professors Elliot Aronson, Tim Wilson, and Robin Akert. Akert interviewed 344 college-age men and women. Not surprisingly, breakees (the ones dumped) were miserable. They reported high levels of loneliness, depression, and anger, plus physical disorders such as headaches, stomach aches, and poor sleep. The breakers (the ones doing the dumping) felt some guilt and unhappiness but had far fewer physical symptoms. The "mutual" role (amicably agreeing to end it) helped individuals evade some of the negatives, with neither partner being as upset or hurt. They didn't get off as easily as breakers, however. Sixty percent of the mutuals reported physical symptoms.

So if you're in a romance your partner seems inclined to break off, try to end it mutually. "Unfortunately for your partner, if you're about to be in the role of breaker, you will experience less pain and suffering if you hold to that role. However, switching from breaker to mutual would be an act of kindness to your soon-to-be ex-loved one."

Does playing "hard to get" work?

Conventional wisdom says "yes." Things you have to work hard for are more appreciated. By definition, you have to work harder for potential dates that play "hard to get." Therefore, potential dates who play "hard to get" are more appreciated.

It sounds reasonable, but it's wrong, say romance researchers. Nobody wants to make a pitch and hear, "Get lost, creep!" In other words, daters don't want to put their fragile hearts and souls on the chopping block. The bottom line, says Michael R. Liebowitz, professor of clinical psychiatry at Columbia University, is that men (the subjects studied) prefer women who are hard to attract to women who are easy, but only if it's not hard for them. "That is, men want someone who is hard for other men, but easy for them, to attract."

Doesn't it just figure?

Do men dream more about their wives, other women, or other men?

Men generally dream more about other men than about women—sexy or not—whereas women split their dreamtime 50-50, said pioneer sleep researcher Calvin Hall.

This difference appears in thousands of dreams reported from around the world, says University of Manitoba psychologist David Koulack. In 29 of the 35 different groups and cultures studied, including U.S. college students, the Zulu, the Hopi, Mt. Everest climbers, Peruvian adolescents, and Guatemalan sleepers, this sex difference showed up.

The obvious explanation is that males may be more important for males than are females. Males spend more significant time together, especially on the job. And so men mull over male-male camaraderies, conflicts, and competitions in their dreams. But as women enter the workforce—which is still male dominated—their dreams swing over to where the action is—women, too, begin to dream more about men.

Does "romance" exist in all cultures?

Something like romantic love seems to exist in all cultures, says Creighton University philosopher Richard J. White. The ancient Egyptians, Hebrews, and Persians all had their own love poetry, and in ancient Athens, love between an older (male) lover and younger (male) beloved was described in rapturous tones.

But not every society seems to approve. The early Christians, like St. Augustine, worried that deep involvement with another person would take away from the love of God. And likewise in the society portrayed in *Romeo and Juliet*, romantic love is viewed as dangerous and a threat to the established order.

About 200 or so years ago, however, in the days of Keats, Byron, Shelley, et al., romantic love started to become the popular phenomenon it is today. "Now there is the common idea that a life without amour is a life that is wasted. It wasn't always that way, and maybe in years to come romantic love will become less important."

Q: Can someone be allergic to kissing?

Yes, says *Science News* magazine. But the offending ingredient here is not lips or saliva. It is food eaten by the other person, often peanuts, says Suzanne S. Teuber of the University of California-Davis School of Medicine. In a number of cases, the kissees had developed hives where they were kissed, even though some of the kissers had eaten the nuts hours earlier. Occasionally a reaction occurred despite a teethbrushing in the interim. Symptoms tended to commence within a minute, such as localized itching and swelling. One serious reaction was in a peanut-allergic boy; another sent a young woman to an emergency room with anaphylaxis after kissing her shrimp-eating boyfriend.

Some 2 to 4 percent of the U.S. population have food allergies—about 1 percent to peanuts or tree nuts such as walnuts, cashews, and almonds, and 2 percent to seafood. In one UC-Davis survey, 6 percent of the allergic participants reported reactions from kissing. Similar numbers came out of a European survey. Scott Sicherer of New York's Mt. Sinai School of Medicine reported that a peck on the cheek is unlikely to cause a severe problem. But longer, more passionate kisses with much saliva exchanged just might.

For people with severe food allergies, advised Sicherer, always carry self-injectable epinephrine. "It will buy time to get to a hospital emergency room."

Why is it so hard to talk to someone you're attracted to?

You're at a party and talking to someone you're attracted to—barely a few feet apart, maybe even nose to nose. You stay like this for three to five minutes (typical for a party chat). The two of you look repeatedly into each other's eyes—the listener looking, the speaker looking away, then back and forth in facial minuet. Here conversation can be a real ordeal, says anthropologist David Givens.

"No other animal hooks up face to face in courtship quite the way we do.... Eye to eye, our heartbeat rate increases. A glance away for only 3 seconds (to relax) can bring the rate down as much as 10 to 15 beats per minute."

You see all the facial signals up close—the hair; the skin's smoothness, wrinkles, pouches, and pits; the teeth; and the whites of the eyes. You smell the breath, feel the body heat, hear the subtlest voice tones.

Two levels operate at the same time: The words spell out the meeting of minds, but beneath is an emotionally loaded, hormonal experience. "A couple can talk about nuclear physics or Malibu—it doesn't matter—and flirt at the same time. But it can all be agony. Speaking doesn't just put you on the line, it hangs you way over the edge."

Q: What was romance like when Shakespeare wrote *Romeo and Juliet*?

A: Sex appeal was likely a rare commodity back then, with most people suffering from poor nutrition and lacking even the energy for sex, says historian Lawrence Stone of Princeton University.

Even if they had the energy, often along with it went lice-filled hair, bad breath, and rotting teeth. People rarely washed, add psychologist Elaine Hatfield and historian Richard Rapson. Their skin crawled with eczema, scabs, running sores, oozing ulcers, and other disfiguring skin diseases.

"Men or women who engaged in sexual relations were likely to catch any number of venereal diseases. For example, James Boswell, the 18th-century biographer, contracted gonorrhea at least 17 times."

DEATH

Part 3

Q: How long can you survive buried alive?

I n horror stories, people often survive for days in airtight coffins underground. But don't believe them, says Jan Bondeson, M.D. of the University of Wales College of Medicine.

Live burials were a morbid curiosity of the 19th century. Some argued that as many as 10 of every 100 burials were premature.

Accounts of "childbirth in a coffin" were taken as a sure sign of horrific endings. But putrefactive gases inside the cadaver would in some cases result in intra-abdominal pressure strong enough to expel an unborn child. A mid-19th-century German article described a case of a "muffled explosion" from the coffin of a pregnant woman interred 24 hours earlier. "The unborn child had been expelled from the womb with considerable force."

For people actually buried alive, one physiologist calculated the air content would not last more than 60 minutes. Then in 1859, a Dr. von Roser put this to a test by burying rats, mice, and a large dog in an airtight coffin with a glass lid. A few rodents stayed alive by eating others and gnawing their way out. The dog, however, grew wretched after two hours, and was dead in three.

Since a dog takes up less space, says Bondeson, the good doctor figured that 60 minutes for a human is about right.

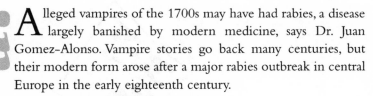

Could vampires have really existed?

A lleged vampires of the 1700s may have had rabies, a disease largely banished by modern medicine, says Dr. Juan Gomez-Alonso. Vampire stories go back many centuries, but their modern form arose after a major rabies outbreak in central Europe in the early eighteenth century.

Here's the evidence:

- Vampires bite people; so do 25 percent of men with rabies.

- Vampires seduce women; rabies patients are often hypersexual, in some cases having sex up to 30 times a day.

- Dracula appeared on moonlit balconies; rabies patients often have disturbed sleep cycles and insomnia.

- Vampires shun mirrors and garlic; rabies patients are hypersensitive to strong stimuli, such as lights, mirrors, and garlicky odors. The mirror connection is so strong, says Gomez-Alonso, that in the past, a man was not considered rabid if he could stand to see his reflection.

- Where there are vampires, there are often wolves and bats—two animals that can contract rabies and pass it on through bloody bites.

Q: Can dead bodies literally spin in their graves?

A: Much folklore over the centuries traces to people's lack of knowledge of the natural processes surrounding corpses, writes Paul Barber of the Fowler Museum of Cultural History, the University of California, Los Angeles:

- The classic theme of a sinner's hand reaching up out of the grave has its origins in common shallow burials of the past and the presence of scavenging animals that latch onto whatever body part can be dragged up, usually an arm.

- Pregnant women who die may expel the fetus a few days later, making it appear they were buried alive and gave birth in the coffin.

- Chewing and suckling sounds are not uncommon in mortuaries and graveyards. "When a corpse bursts as a result of bloating, the emission of gases, body fluids, and maggots—present in astonishing numbers—may be audible." One 19th-century gravedigger said he heard ghastly noises for three weeks following a shallow burial.

- The limbs of a lifted corpse can react in frightening ways once rigor mortis passes and they're free to "flail about" under the force of gravity.

- A body may "cry out" when a stake is driven into the chest, not because it is a vampire coming to life but because this forces a rush of air past the glottis.

Could you die from having sex in a dream?

Dream sex is associated with physiological changes very similar to corresponding activities in the waking state, says Stanford research associate Stephen LaBerge, Ph.D. But an important exception is heart rate, which increases only slightly in dreams compared to doubling or tripling during actual sex.

This fact may have practical benefit. For heart patients, sex can be a dangerous and sometimes fatal form of exercise. "Dream sex, in contrast, appears to be completely safe for everyone, and for many paralyzed people, it may be the only form of sexual release available."

Will AC or DC electricity kill you faster?

This debate raged as part of the 1880s "battle of the currents," with Thomas Edison on one side boosting his new first DC power plant, and engineer William Stanley his even newer AC power plant, says James D. Livingston of the Massachusetts Institute of Technology.

AC's big advantage is its transformability into higher or lower voltages, but it also got the rap of being more dangerous. Safety critic Harold Brown did a demo where a dog survived 1,000 volts DC, but died when subjected to 300 volts AC. "But that's a small dog, not a human," AC boosters countered. Brown answered by stepping up his demos to include more human-sized calves and even a horse.

Bring on George Westinghouse, who still maintained that AC wasn't more dangerous. Brown challenged him to a duel by electricity: "Brown would take DC through his body and Westinghouse would take AC, at gradually increasing voltages until one cried 'enough' and admitted he was wrong (or died and proved it!). Westinghouse declined the offer."

Watching all this were New York State officials, who soon introduced the first electric chair—an AC chair.

Actually, says Livingston, both types of current can be deadly depending on how the power source is constructed and how the power is delivered to the body.

Does an MRI of a dead body look different from one of a living body?

If a "fresh" body were found, say, on its way to the morgue, the differences wouldn't be that obvious, says University of Wisconsin medical physicist emeritus James A. Sorenson. This assumes that the corpse has no missing parts, no strange-looking fluid pockets, and no chemical preservatives pumped in yet that might make some tissues show up abnormally bright or not at all.

Water after cell death gets redistributed, causing changes in tissue contrasts. For instance, the contrast between white matter and gray matter of the brain would be suppressed after cell death and deterioration. "An experienced technician might notice these changes. But they would be subtle."

"Dead or alive" would be easier to tell if you looked at the images side-by-side. The body movements or pulsatile flow of a living body could show up as "ghost images." Here the moving structures appear displaced from their true body location.

It is virtually impossible to remain still for an entire MRI scan, and, of course, the heart and blood will move regardless, says Sorenson. So researchers have worked hard to suppress ghosts, since they interfere with image interpretation. "Still, I think an experienced observer on careful inspection wouldn't be fooled, except perhaps on images with no prominent arteries, such as the hands and feet. In a vascular scan designed to show only the blood vessels, there would be nothing to see, because there would be no moving blood to create an image. That would be a 'dead giveaway.'"

Do patients sometimes wake up in the middle of surgery?

The anesthetic wears off. The patient wakes up just in time to feel the scalpel sinking in. But he or she can't cry out to halt the procedure....

This situation is a half truth, say University of Washington anesthesiologist Christopher Bernards and neuroscientist Eric H. Chudler. Awareness during surgery, though very rare, does occur. The patient may not tolerate the anesthetic well, so the anesthesiologist keeps the amount down to avoid depressing the blood pressure to dangerous levels. Or the patient may have a history of alcoholism or abuse of valium or sleeping pills, making the brain resistant to the sedative effect of the anesthetic.

The patients usually say that pain wasn't a problem. But the extreme anxiety due to being unable to move was. Thank goodness for the standard muscle relaxant that induces temporary "paralysis," keeping bodily movements in check, say Bernards and Chudler. This same relaxant also makes muscles softer and easier to spread, such as abdominal muscles that are easily parted to expose the site of surgical interest.

Q: Can a voodoo curse kill?

A: Psychologist and author Dennis Coon wrote of a terrified young woman admitted to a hospital because she believed she was going to die. A midwife had predicted that the woman's two sisters would die on their 16th and 21st birthdays, and that the woman herself would die on her 23rd birthday.

Her sisters died as predicted, and now it was three days before her 23rd birthday. The following day the woman was found dead in her hospital bed, "an apparent victim of her own terror."

Such victims may die in one of two ways, says Coon:

Physiologically, the intense arousal causes a sharp rise in blood sugar, the heart beats faster, digestion slows or stops, and blood flow to the skin is reduced. These "fight-or-flight" reactions generally increase the chances of survival in an emergency, but they can kill an older person or someone in bad health.

And if the initial emotion doesn't prove fatal, "parasympathetic rebound" might: Following heightened arousal, the body works to calm all the accelerated processes and, in doing so, may go too far: Even in a young, vigorous person, the counter-slow-down may actually stop the heart.

Thus voodoo, like all terrors, can get you coming or going.

Q: Can you be in a terrible accident without feeling pain?

A: In 1857, Scottish explorer David Livingstone recounted being attacked by a big cat:

"Growling horribly close to my ear, he shook me as a terrier does a rat. The shock produced a stupor similar to that which seems to be felt by a mouse after the first shake of the cat. It caused a sort of dreaminess, in which there was no sense of pain nor feeling of terror, though [I was] quite conscious of all that was happening."

Then a friend fired a gun and scared the lion off, saving the badly injured Livingstone.

Eminent doctor and author Lewis Thomas described witnessing a jeep accident during World War II where two young MPs were trapped inside the crushed steel, both mortally injured. "We had a conversation while people with the right tools were prying them free. Sorry about the accident, they said. No, they said, they felt fine. Is everyone else okay, one of them said. Well, the other one said, no hurry now. And then they died."

Such stories are common where people encounter violent life-threatening circumstances. In many grave situations where pain would seem to serve no biological purpose, the brain apparently shuts it off with natural stress chemicals called endorphins (for "endogenous morphine").

Q: What would it be like to die in outer space?

A: In the movie *Mission to Mars*, an astronaut dies in outer space. He tries to grab a passing satellite but his hook misses. He's stranded and doomed since his compatriots lack fuel for a rescue. To prevent his fellow astronaut-wife from dying by trying to save him, he commits suicide by yanking off his helmet. "Instant iceball, with his face becoming distorted and icky!" is how Sonoma State University astronomer Philip Plait described that moment in the newspaper *Frankfurter Allgemeine Zeitung*.

But space isn't really cold because it has almost no matter to be cold. Stuff on Earth cools largely by conduction (heat transfer through touch) or convection (when hot things rise and cool things sink), but not in space where there's no medium like air or water. Radiation cooling does operate in space, but slowly.

Small consolation. And it only gets worse. From NASA's *Bioastronautics Data Book*, an astronaut exposed to a vacuum has maybe 10 seconds before losing consciousness, after which he or she becomes paralyzed, then convulses, and then is paralyzed again. The sudden pressure drop would explode air out of the lungs, rupturing tissue and damaging ears.

In lowered pressure, water vapor streams from the mouth and nose—a cooling process—icing these body parts. "But this would take a minute or so, and the body would not freeze solid," says Plait. Beyond this, don't even ask.

Would a penny tossed off the Empire State Building bore through a human skull?

Ignoring air resistance, a penny falling 1,250 feet would reach a speed of 193 mph by the time it hit the street. That's much less than a bullet's speed, but is still dangerous, says Louis A. Bloomfield, professor of physics at the University of Virginia.

A falling object hits a "terminal velocity" when air drag balances weight. At this point, no further acceleration occurs. For an aerodynamically poor penny, this would be early in its descent. Its speed would probably peak at less than 100 mph. So if you could see the penny coming, you might catch it in your hand and suffer little more than a bruise.

A dropped object to worry about would be a ballpoint pen. It's heavier, and instead of tumbling like a penny it pierces through the air more like a falling arrow—it's faster and deadlier. Not at all theoretical is what happens to a bullet fired up into the air, which then falls thousands of feet. "This can definitely kill," says Bloomfield, "and a relative of mine died this way at a Mardi Gras years ago."

Q: Would a guillotined head feel itself hit the ground?

In France in the days of the blade, some of the condemned were asked to blink their eyes to show continued consciousness after decapitation, and a few heads blinked for up to 30 seconds, says Dale McIntyre in *New Scientist*. "How much of this was voluntary and how much due to nerve reflex action is speculation. Most nations with science sophisticated enough to determine this question have long since abandoned decapitation as a legal tool."

Addressing the reflex issue, one Dr. Beaurieux observed the execution of a murderer in 1905, says writer Alister Kershaw in his history of the guillotine. First he saw spasmodic movements of eyes and lips for five to six seconds. Then the face relaxed, the lids half closed, "exactly as in the dying whom we have occasion to observe every day in the exercise of our profession."

"It was then that I called in a strong sharp voice:'Languille!'" The lids lifted, and Languille's "undeniably living eyes" fixed on the doctor, after which they closed again. Moments later he called out again, fetching another look by Languille. But a third call went unheeded.

"I have just recounted to you ... what I was able to observe. The whole thing had lasted 25 to 30 seconds."

Q: Will a household cat or dog eat its owner who dies, leaving the pet alone and unfed?

A: Pets feed on their deceased owners from time to time, says North Carolina Chief Medical Examiner John D. Butts, M.D. This happens rarely, though—most likely because the opportunity does not occur that often. "In our experience, the animal is usually a dog."

Dogs will target certain parts of the human body first, says State University of New York-Canton criminal investigations and forensic science professor Steve Gilbert, MFS, ABD. First comes the daily licking of the deceased's face and hands. Then the dog eats away at these parts, and elsewhere. "There is no difference between dogs and cats when it comes to survival: Both will eat what they must."

Far grimmer than the cases of dead owners, as reported in *The American Journal of Forensic Medicine and Pathology*, are the cases of elderly people who signal weakness and vulnerability to groups of predatory dogs.

Is it common to die on the toilet?

It's more common to die in bed, since people spend about eight hours a day there. (Dreaming is actually the riskiest part of sleeping.) Baths, too, take their steamy, slippery toll.

By one tally of sudden deaths in Osaka, Japan, in 8 percent of cases the fatal symptoms began in the lavatory, 17 percent in the bath, and 31 percent in bed, says Adam Hart-Davis, science writer and member of the British Toilet Association.

Elvis Presley, the undisputed "king," reportedly died on the throne. More shocking was a guy in Ryde, on the Isle of Wight, who sat on a metal lavatory seat and was electrocuted by a faulty cable.

Then there are the streetpeople of New York City for whom their pissoir is a subway station. Waking up, writes Hart-Davis, they'll "wander over to the edge of the platform, and unthinkingly pee on the live rail. Urine is a solution of salts in water, so it's a good conductor of electricity."

Can a person die laughing?

In then–Tanganyika (now Tanzania) in 1962, a group of schoolgirls started laughing so hard they cried, and the laughter spread to nearby villages. This happened in such epidemic proportions schools were forced to close for six months, says neuroscientist and laughter specialist Robert Provine.

The British journal *Nature* told of a 15-year-old epileptic undergoing brain surgery. She was kept awake so she could give feedback. Suddenly she began laughing uncontrollably, telling the doctors, "You guys are just so funny."

One morning after a violent headache, 58-year-old Ruth was seized by a laughing fit that went on for hours in spite of a morphine injection, reports neuroscientist V. S. Ramachandran. She became exhausted, the laughter persisting as noiseless grimaces, until she utterly collapsed. "I can say that she literally died laughing."

The few reported cases of pathological laughter all seem to involve parts of the limbic system affecting emotions, says Ramachandran. "Given the complexity of laughter and its infinite cultural overtones, I find it intriguing that a small cluster of brain structures is behind the phenomenon—a sort of 'laughter circuit.'"

Can dreams serve as early warning signs of illness?

This ancient notion was embraced by Aristotle and Hippocrates, father of Greek medicine, says internationally recognized dream authority Dr. Robert L. Van de Castle. Second-century Greek physician Galen told of a man who dreamed his leg turned to stone, then days later suffered paralysis.

Bestselling author Dr. Bernard Siegel described a journalist's dream where torturers placed hot coals under the dreamer's chin. Later, doctors pooh-poohed the man's fear of cancer, until tests confirmed cancer of the thyroid.

"Dreams can be like X-rays in some cases," says Van de Castle. One woman dreamed repeatedly that her leg was being examined by a nurse who would hold a candle closer and closer until it started to burn her shin. Soon after the dream series, the woman was diagnosed with a bone infection.

Neurologist Daniel Schneider recounts the case of a heavy smoker who dreamed he was in an army combat zone trying to take cover in the hollow of a tree, but enemy fire cut him in half across the chest. A checkup revealed a small lung tumor that had not yet metastasized.

A depressed patient of Schneider's dreamed he was onstage, opened a violin case, pulled out a machine gun, and shot himself. As the rat-a-tat-tat of the gun raged, he awoke, only to suffer a heart attack minutes later. The loud, rhythmic rat-a-tat-tat heard in his dream was likely his own heart racing.

This is a fertile area for more research, says Van de Castle. "Many dreams seem to have a warning function, and may be useful or adaptive if heeded."

If you die in a dream, will you actually die in your sleep?

There are many forms of death in dreams, and the dreamers live to tell about it the next day, says Dr. Antonio Zadra, of the Dream and Nightmare Laboratory, Montreal Sacre-Coeur Hospital.

There are even attempted suicides in dreams. Some lucid dreamers, who become aware they're in the midst of a dream, have experimented with killing themselves to see what would happen. "As far as I know, none succeeded. What is frequently reported is that something goes wrong." One person tried to shoot himself in the head but the gun would only fire if it was pointed away. Another tried to jump off a cliff but ended up floating down safely.

"But I'd also like to add that if you happen to be dreaming when you really die, I would venture to say that you'll die in your dream as well."

What food has made the most people sick?

On TV's *Survivor*, many of the contestants gag or vomit after eating insects or eyeballs or drinking blood. But notice how others just toss these down and smile, says University of California–Davis nutritionist Louis Grivetti. There are two types of food sickness: You can get ill from something culturally inappropriate, or from something containing a pathogen or toxin.

Mole dishes sold at the great plaza in downtown Mexico City are a delight to the senses, with their wonderful chocolate sauces (containing from 10 to 50 ingredients). But for many gringos, they spell instant "death." Food sitting too long on a hot summer day may be deadly for a very different reason.

"If I were to choose just *one* food, it would be common drinking water—worldwide some 40,000 children per day die of diarrhea-related diseases primarily from unsanitary water," says Grivetti.

Biologist Grady W. Chism, III, of Indiana University-Purdue University Indianapolis, agrees about water, but for a food chooses potatoes. Properly stored, they're no problem. But exposed to sunlight, they produce solanine. Folklore says potatoes in open barrels on the sides of Conestoga wagons may have caused the deaths of more U.S. settlers than arrows and bullets. "Allergens are another, possibly causing headaches, rashes...and an estimated 150 deaths a year in the U.S. alone."

Shellfish and fish—especially raw—just might be it, says Don Schaffner of Rutgers University. "Sushi would be my guess."

Is it possible to be scared to death?

In 1936, in India, recounts Nobel Laureate and physician Bernard Lown, an astonishing experiment was conducted on a prisoner condemned to die by hanging. He was given the choice instead of being "exsanguinated," or having his blood let out, because this would be gradual and relatively painless. The victim agreed, and was strapped to the bed and blindfolded.

Unbeknownst to him, water containers were attached to the four bedposts and drip buckets set up below. Then after light scratches were made on his four extremities, the fake drip brigade began: First rapidly, then slowly, always loudly. "As the dripping of water stopped, the healthy young man's heart stopped also. He was dead, having lost not a drop of blood."

Dying of fright can occur in one of two ways, explains psychologist Dennis Coon. The stepped-up heartbeat and other physiological reactions of the "fight-or-flight" response can kill directly; or "parasympathetic rebound" can be deadly, where the body works to calm itself and goes too far the other way, and actually stops the heart. Cases abound of soldiers dying of fright in savage battles, or of people dying at other very emotional times. Voodoo deaths also pay testament to the amazing power of the mind over the body.

Could a black hole kill us all?

There are three possible scenarios here, says North Carolina State University physicist John Blondin:

(1) Star-sized black hole: These are known to exist, with more than a dozen already identified in our galaxy. There could be *many* more wandering through space. But it's hard to say because they're only visible when a normal star is close by, dumping gas into the black hole, making it glow. "If one of these wandered by, we could kiss our *** goodbye, and it would not be pleasant. There'd be ample warning, months or even years, because their intense gravity would distort the night sky star patterns, but *nothing* we could do. The end would be gruesome as the tidal pull of the black hole ripped Earth into pieces and sucked them and us into darkness."

(2) Supermassive black hole: These have the mass of millions or billions of stars and exist at the cores of galaxies, including the Milky Way. There's no chance of being sucked into one of these in the next billion years. Here the tidal pull is weak, so you could fall in completely intact. Then what? A new dimension of space-time?

(3) Mini black hole: These have never been seen but are possible. Small as an asteroid, a mini black hole would probably fly straight through Earth, "eating" its way through. Because of its small size, we wouldn't see it coming. The worst case would be if one was big enough to swallow enough mass to slow down. It could then start sucking up the Earth bit by bit from the inside out.

Is it unreasonable to be afraid of being buried alive?

The ancient Greeks and Romans waited three days or more for putrefaction to begin before burying or cremating, says Kenneth Iserson, M.D. Before consigning a body to the pyre, the Romans called out the person's name three times and cut off a finger to see if it bled. In Shakespeare's day, a feather or mirror was held near the mouth to look for signs of breath.

In 1864, reported Dr. Franz Hartmann, a New York physician began a "post-mortem" with a cut across a man's chest, whereupon the guy jumped up and grabbed the doctor by the throat. The examiner died on the spot of apoplexy, while the "deceased" made a full recovery.

Fewer mistakes are made today, thanks to medical technology. But as recently as 1993, a 40-year-old New York woman revived two hours after being pronounced dead by both an ambulance crew and the coroner's office. She had been so cold from outdoor exposure, it was claimed, that no heart beat, blood pressure, or breathing could be detected.

Mistakes notwithstanding, the fear of being buried alive "should have vanished with the introduction of arterial embalming in the 1880s and 1890s," says Iserson. "Once the chemicals are injected into bodies, people are dead, whether or not they had been originally."

ANIMALS

Part 4

Q: Does a "dog year" really equal seven human years?

a: Nope, says Michigan State University veterinary medicine student Brian Dawson. A one-year-old dog has reached sexual maturity, but even a precocious seven-year-old human has a way to go. At the other end of the scale, the world's oldest dog at 29 would be 203!

The formula breaks down, especially at the extremes. "Since dogs' lifespans are usually limited to 15 to 20 years, it should be enough to say that a 2-year-old dog is young, a 9-year-old is middle-aged, and a 14-year-old is old."

If you must apply a formula, try this: Count dog year No. 1 as 15 human years, year No. 2 as 10 more, and every year after that as an additional 3. This way, a 10-year-old dog corresponds more realistically with a 49-year-old human, not a 70-year-old.

How is animal sex different from human sex?

Forget the fringe stuff—the mainstream stuff is weird in comparison.

For starters, we humans don't do it like other primates, preferring to keep our lovemaking private, says University of California physiologist Jared Diamond. That's pretty strange for a "social" species.

Moreover, the woman's infertility—whether due to being pregnant, post-menopausal, or the time of the month—doesn't stop her or him. The fact is, she rarely has a good idea if she's fertile at any particular time, and certainly doesn't advertise it through distinctive smells, colors, and sounds as do other mammals. Then there's menopause: Virtually no other mammal has this, says Diamond.

And as for the male, what's the point of such an unnecessarily large apparatus, exceeding even that of a gorilla?

Finally, why do men so often stay with the women they impregnate to help raise the children? In this, too, we are unique among mammals—and uniquely lucky, many proud Dads might add.

Q: If a poisonous snake bites itself, will it die?

It depends on the venom, how much is injected (often none is), and the health of the snake, says University of Florida herpetologist Max Nickerson. The self-inflicted bite could cause little reaction, or death if vital tissue is involved.

Why would a snake bite itself? Accidents happen, as when a strike badly misses its target. Then there are those stories of cornered or injured snakes biting themselves to commit suicide. Certainly a snake under stress may self-inflict, most notably when it's pinned with a hook or stick and goes into a frenzy. Bright lights and heat may also trigger it: Brought into a TV studio, a sidewinder rattlesnake began biting itself and died shortly afterward.

Then there's the cold factor: Lowering reptiles' body temperature to freezing has been used as a humane way to euthanize them. "On several instances, I found rattlesnakes with their fangs embedded mid-body after euthanization."

Can animals be hypnotized?

If you rub the abdomen of a rabbit or hold a chicken on its back and cover its eyes for a minute, it will lie still for some time, says Cornell animal behavior veterinarian Katherine Houpt. Or you can swing the chicken back and forth with its head beneath its wing. That state of lying still is called "tonic immobility."

Other techniques seem to bring on a trance-like state, such as stroking the tentacles of an octopus or the stomach of an alligator or crocodile—if you have the courage, adds veterinary anatomist Michael Bryden of the University of Sydney, New South Wales. "In each case, the animal might just lie motionless, and permit simple procedures to be done on it."

Fishermen might grab the tip of a shark's tail and bend it over, rendering the accidentally netted catch "immediately unresponsive, almost catatonic for 30 to 90 seconds," reports *Diver* magazine. The hook can then be removed and the shark let free—and saved.

Is any of this really hypnosis? "Tough to say," answers Ohio State veterinary surgeon David Anderson. Animals can't relate their experiences to us, at least not in a way we can understand. "I've observed alpacas and llamas 'calmed' into a state of relaxation by gently rubbing the upper gum just beneath the cleft in the upper lip. The animals stop resisting being held, and stop vocalizing."

"But I have not observed any results from suggestions I have made to them while they are in this state!"

When lightning strikes a body of water, do all the fish in the water die?

It's no different for fish than for human swimmers who foolishly hang around in a storm. Water is a pretty good electrical conductor.

Lightning is high in voltage (100 million to a billion volts) and in current, but the kill-charge dissipates fast beneath the surface of the water away from the strike point. "So fish do get killed, but probably not more than 30 yards away in fresh water, 10 yards in salt water, though good information is scarce," says University of Florida engineer Martin Uman.

Fish also tend to hang out well below the surface, especially at night. This makes them less vulnerable to lightning.

"In my own experience, I have only heard of one case—goldfish in a pond being struck dead by lightning, back in the '70s," says Paul Skelton, director of the JLB Smith Institute of Ichthyology, South Africa. Scuba-diving Roger Bills—another researcher there—tells of being shocked several times in Lake Tanganyika. "On one dive, at around 20 meters I didn't feel a thing, though I saw flashes going off all the time. During decompression was when I got shocked, worse as I neared the surface—like a cattle electric fence, not incapacitating but very unpleasant. I have no idea how far off the lightning was, but I never saw any dead fish."

Q: Can humans and chimps interbreed?

No one knows—or if they do, they're keeping it to themselves, says Dr. Ian York of the University of Massachusetts Medical Center.

Let's start with a relatively simple question: Can humans and chimps interbreed to produce fertile offspring? The answer is a pretty clear "No." Chimps and humans have different numbers of chromosomes—humans have 23 pairs, chimps have 24. This means that their hypothetical offspring would most likely be sterile.

But members of different species can sometimes interbreed. The most obvious examples are horses and donkeys. They have different chromosome numbers but can interbreed like nobody's business to produce mules or hinnies.

Beyond this, though, there's no simple rule for determining which species can interbreed. Humans and chimps diverged at roughly the same time as did horses and donkeys (between 5 and 10 million years ago). Very roughly, humans and chimps are about as genetically similar as are horses and donkeys. But that doesn't guarantee anything; it's a case-by-case matter. Some species that are more similar than humans and chimps cannot interbreed.

"In principle, it's entirely possible chimps and humans can interbreed. In principle, it's entirely possible that they can't. Unless the experiment is done, we won't know," says York.

Can a dog tell if its owner is sick?

D on't sell Rover short, for along with cases of canines wresting their owners out of bed just before an earthquake or tornado strikes are the many documented instances of dogs warning their special people of an imminent medical problem.

Author Rupert Sheldrake, Ph.D., tells of a collie named Molly that senses her owner Lise's epileptic seizures half an hour in advance, then starts staring, barking, and licking to alert her. Whether Molly is sensing subtle muscular tremors, electrical disturbances, a distinctive odor, or something else, this canine capability underpins organizations such as Support Dogs and the Seizure Alert Dog Association.

The journal *Diabetic Medicine* reported on a study of 43 pet-owning patients, 14 of whom said their dogs could sense episodes of low blood sugar—possibly based on body odor.

Then there's the case of Marilyn, whose Shetland sheepdog Tricia took to sniffing and nuzzling an area of the woman's back around a small mole, reports psychologist Stanley Coren. The dog grew more and more insistent, until Marilyn finally went to her doctor. She was stunned to get a diagnosis of melanoma. "Tricia's early warning probably saved Marilyn's life."

As a researcher told Coren, "Inspection by a dog may one day become a routine part of cancer screening."

Could a human outrun a dinosaur?

By analyzing footprint remains and estimated weights of large dinos, zoologist R. McNeill Alexander concludes that a 37-ton brontosaur may well have been about as athletic as an elephant. Elephants can't gallop or jump but can run at least 10 mph.

The much lighter, horned Triceratops, at 7 tons, may have been able to gallop, but probably not as fast as a buffalo, speculates Alexander.

A really fast human can run at about 27 mph very briefly, and 22 mph averaged over a 100 meter race. But most of us run a lot slower than this, so fleeing a pursuing dinosaur would have been an iffy proposition, depending on one's physical condition, the type of dinosaur and terrain, and whether the beast recently had lunch.

Do seeing-eye dogs have 20/20 vision?

Dogs typically have only 20/75 vision, but German shepherds and Labradors in guide-dog programs are bred for keener eyesight and probably come closer to 20/20, says Ralph Hamor, veterinary ophthalmologist at the University of Illinois.

Still, like other dogs, they're color blind, unable to distinguish among green, yellow, orange, and red. At a traffic light, they must draw from on-signal position and traffic noise and flow to determine if it's a "go."

Yet the same wide set of canine eyes that weakens acuity enhances peripheral vision, so a dog can see both up and down the street while looking straight ahead.

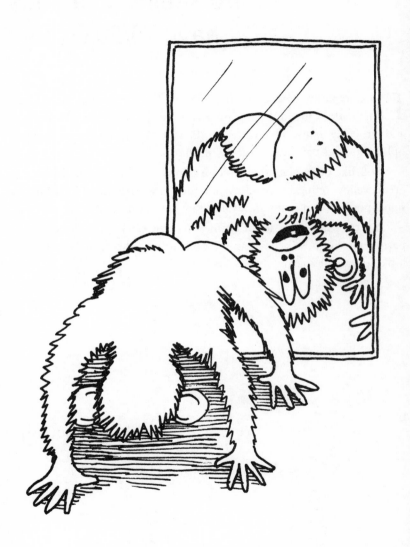

Q: Do dogs watch TV?

When the magician and skeptic James "Amazing" Randi, who funds a $1,000,000 prize for anyone who has evidence of a paranormal event, got wind of a claim that a poodle named Terr was a big TV watcher, he decided to investigate.

First he exposed the dog to neutral non-canine videos, then to ones where Lassie came in after three minutes or so. As Randi observed, Terr quickly lost interest when people were depicted, but when Lassie appeared Terr immediately perked up and began barking.

"Terr would never bite at the image of a person, but only the dog," reported fellow-skeptic Al Seckel in *Laser* magazine. "Terr would also follow the dog's image across the screen. Terr was able to distinguish a running Lassie when the image was only an inch across. And Terr went positively berserk when a cat appeared on the screen!"

Terr, it turned out, wasn't fond of cartoons but loved nature shows, especially those featuring fellow four-leggers, watching them for up to an hour.

Said an amused Randi: "I wouldn't have believed it! This dog actually watches TV!"

Do animals like alcohol?

Here's some evidence from Ronald K. Siegel of the UCLA School of Medicine:

- Bumblebees, hornets, and wasps will lose their coordination and become temporarily grounded after ingesting fermented fruits.

- In earlier times, illegal stills were sometimes traced by the trail of tipsy livestock that had sniffed out the mash.

- African villagers have been known to rid themselves of rodents by leaving out bowls of milk and beer, then rounding up the woozy pests in the morning.

- Parrots are said to become more talkative after eating fermented fruits or sipping alcoholic beverages. "The birds stop talking and drinking only after they fall over," says Siegel. One dealer in rare birds overdosed his parrots on tequila to quiet them for smuggling across the Mexican border.

- Herd-feeding elephants, afraid they won't get their fair share of fermented fruit, will at times gorge themselves into swaying inebriation.

- One dog that lived near a brewery would often drink beer instead of eating.

- If an experimental "bar" is opened for monkeys on a 24-hour basis, they'll go into binge and abstinence cycles, much like addicted people.

Why do dogs make such great pets?

Tens of thousands of years ago, wild dogs—wolves really—hung out around human campsites. They fed on scraps, and barked out warnings of approaching strangers in return, says writer Guy Murchie. Later, "domesticated" dogs joined humans in the hunt, serving as finders, retrievers, protectors, and companions.

Today, by the tens of millions, we live under the same roof. Fundamentally, our mental makeups mesh. As wolf descendants, dogs are pack animals with complicated social patterns, facial signalings, and a strong hierarchy of dominance and submission, says archaeozoologist Juliet Clutton-Brock of the Natural History Museum, London. Adopt a dog and you become pack "leader."

Out of the complex give-and-take of human-dog cohabitation, there grows empathy so close that a puppy in a smile-a-lot family "will actually mimic this expression by a sideways grin of the lips and muscles around the mouth."

Hundreds of generations of selective breeding have remade dogs closer to our heart's desire—or, rather, desires, as 400 to 800 different breeds attest. For example, humans have bred dogs for tractability and/or floppy ears, while screening out fear of strangers and unfamiliar situations. "No wonder dogs seem so perfectly matched to humanity's requirements and so perfectly adapted to our lives," says psychologist Stanley Coren. "We created them to be so."

Does a dog or a human have the stronger bite?

Bite forces have been actively studied in a number of species, living and extinct, says Jerry D. Harris, director of paleontology at Dixie State College. An American alligator bites with a force of about 19,000 Newtons (4,300 pounds), a hyena about 9,000, a lion 4,200, a dusky shark 3,000, and a Labrador retriever 1,100 Newtons (250 pounds). A human, with a bite force of 1,500 Newtons, actually has a stronger bite than a Labrador retriever (a "strongly built, medium-sized" dog, according the American Kennel Club).

But these force values are for the entire mouth, based on jaw and teeth shape. So while a human beats out the Lab, says Harris, the dog has much pointier teeth distributed in a narrower mouth, making for a more concentrated chomp at the teeth tips. Thus, Duke can do far more damage, though human bites can be plenty nasty.

Q: What's the biggest ever of all Earth's creatures?

A: The mammoth blue whale at around 120 tons, equaling the weight of 1,600 150-pound people, or 120 midsize cars, or 20 African elephants.

It would have taken two large Brachiosaurus dinosaurs, at 50-plus tons each, to rival one whale—whose tongue alone weighs as much as an elephant and heart as much as a car. Whales are seafarers, able to haul around so much mass because the buoyancy force of water largely negates gravity. In fact, a beached whale rests so heavily on its lungs, it is slowly strangled to death.

How do snakes swallow large animals whole?

Needle-like teeth in elastic-muscled, wide-hingeing jaws fix the prey in place, as saliva greases it for the big swallow, says Curator of Herpetology F. Wayne King of the Florida Museum of Natural History.

A python downing a gazelle starts at its nose. As the gazelle is engulfed, its legs fold naturally and lay along its sides, a "bolus bullet." It may take several minutes for the food to pass from mouth to throat to gut. The gut can expand as needed—the python has ribs, but no rib cage. A special trachea hookup allows the snake to pause, take a breath or two, and return to swallowing without choking.

A meal the size of a gazelle might take a week or more to digest, with the snake sinking into a low-metabolic torpor. It wouldn't need to eat again for months, says King.

Can you give sick dogs and cats human medicine?

Many medicines have the same active ingredients across species, but often under a different name, says Texas A&M University veterinary medicine specialist Dr. Deborah Kochevar. In some cases, a vet may use the human product because an appropriate pet product is not available.

Many human conditions, like diabetes, arthritis, and allergies, are also common in pets, though atherosclerosis and heart attack are exceptions. "But dogs and cats are *not* just like little humans, and you should always consult a veterinarian before medicating any animal."

For one thing, doses may differ. Cats, for example, lack a liver enzyme for helping clear aspirin from the bloodstream, so they can only be given small amounts once every two to three days. Tylenol may outright kill them.

And on the psyche side, many people with compulsive behaviors, such as hand washings every 15 minutes, get depressed from the lifestyle disruption, says psychologist John Dworetzky. In the 1970s, the anti-depressive drug clomipramine was found to help these depressions, and then strangely, often the behaviors themselves disappeared.

Later, Judith Rapoport looked at "acral lick dermatitis" in dogs, where they lick their paws so much the skin wears away and bones may show through. The usual treatment is to wrap the paws in bandages, says Dworetzky, but today "something else also works—clomipramine!"

Are there "stealth" bugs?

From the fossil record, we know that bats have been using sonar stalking for at least 50 million years. So bugs have had time to evolve defenses, says Dartmouth's Howard C. Hughes.

Certain moths can detect when a sonar beam has found them, then off they fly at a sharp right angle to the approaching bat. If a similar maneuver doesn't succeed for a green lacewing, it "freezes" its wings and drops straight from the sky. These defenses seem to work, says Hughes, as zoologists estimate that insects with sophisticated sonar detection are only half as likely to fall prey to bats.

Jamming the bat's receiver is another option, i.e., adding confusion with counter-signals. Or some un-tasty insects may use ultrasonics to advertise their presence so as not to be mistaken for a more desirable morsel.

It's too bad that bugs haven't hit upon a form of "stealth" technology, which has rendered military aircraft "invisible" not only to radar but to biosonar as well. So invisible, in fact, that the "U.S. Air Force discovered stealth fighters left outside overnight were often littered with injured or dead bats that had apparently flown into them during the night!"

Not Superstitious   Superstitious

Why is there more superstition about cats than dogs?

The human/cat symbiosis is ancient and deep. Humans provided home and hearth, and the felines provided varmint clearance and cuddly comfort, says Juliet Clutton-Brock of the Natural History Museum, London. Cats have changed little in thousands of years—unlike hyper-bred dogs—with ferals and housecats still closely akin.

Cats are creatures of the night, solitary hunters off in the Black Deep, capable of issuing "the most blood-curdling cries when [they are] fighting." Little wonder cats became associated with witchcraft, sorcery, and sympathetic magic. As late as the 1700s in Britain, superstition held that a cat corpse in the walls of a building would keep it free of rats and mice. Dried cats have been unearthed during old renovations, apparently killed and placed "in a life-like position, sometimes with a rat beside it or in its jaws."

Ancient Egyptians mummified "sacred" cats in enormous numbers. It was forbidden to kill one, and Herodotus tells how if a cat died in the house, the family shaved their eyebrows. But X-ray studies by the Natural History Museum showed many cats were killed as young kits, their necks broken, and possibly sold as votive offerings.

So many mummies were found that 20th-century excavators spread them as fertilizer, says Clutton-Brock, including 19 tons shipped to England to be ground up for the purpose.

How do dolphins see and then tell their "friends" about the sights beheld?

Just as physicians use high-frequency sound (ultrasound) to "see" inside the body without X-rays, and bats emit ultrasonic squeaks to locate objects by their echoes, dolphins can do both of these tricks and a lot more, says physicist Paul Hewitt. But whereas sound is a passive sense for us humans, for them it is the primary sense as they send out sounds and then perceive their surroundings on the basis of the echoes that come back.

Perceive what? The ultrasonic waves emitted enable dolphins to "see" through the bodies of other animals and people. Skin, muscle, and fat are almost transparent, yielding only a thin outline of the body, though bones, teeth, and gas-filled cavities are clearly apparent. "Physical evidence of cancers, tumors, heart attacks and even emotional states can all be 'seen' by the dolphin—as humans have only recently been able to do with ultrasound."

Amazingly, dolphins probably communicate their experience by transmitting the full acoustic image, placing it directly in the minds of other dolphins, speculates Hewitt. No need for a word or symbol for "fish," for example, but just the image of the real thing, just as we might communicate a musical concert via various means of sound reproduction. "Small wonder that the language of the dolphin is very unlike our own!"

Is it only mammals that fart (oops, flatulate)?

At least some cold-blooded animals *do* "pass gas" from the vent/cloaca/rectum, though the sounds won't be like a mammal breaking wind, says Barbara Shields, Ph.D., of the Department of Fisheries and Wildlife of Oregon State University. But it can certainly smell just as bad, or worse!

In many species, this is part of the fecal evacuation process as ingested air escapes with the voiding. Pet fish fed dried pellets often pass a lot of gas. Digestive breakdown of foods by gut bacteria can result in the very smelly flatulence of reptiles, especially snakes.

"My boa and python were the worst, especially after a high-fat meal," says Shields. "There is usually a warning gurgling noise prior to release. In fact, my 22-year-old bullsnake Snippy is gurgling away right now on the 10 mice I fed her last week. A large buildup of gas can result in literally explosive defecation amid a blast of evil wind. I've moved her outside in anticipation of the coming event."

Most noteworthy, however, are fishes in the herring family (Clupeidae) that are able to expel air from the swim bladder through a unique rear duct. Via muscular alteration of duct size and anal aperture, they can control the size of the bubbles streaming out, and hence, control the pitch. "In essence, these fish 'fart a tune,' used to confuse predators and communicate with conspecifics."

Q: What's the strongest material made by an animal?

A: Spider silk. Early on, silkworms provided tough materials for parachutes, pantaloons, and the like. But recent research indicates that spider silk is superior for tasks such as warding off projectiles. "Dragline silk is the strongest animal-stuff there is," says University of Wyoming molecular biologist Randy Lewis. These are the threads spiders dangle from, and use in their outer-web framework. (The inner web is cushier, to snare prey.)

Dragline silk is at least three times stronger than Kevlar, which is used in bulletproof vests, says Lewis. A 1-inch diameter "rope" of this silk, if it could be woven using millions and millions of strands, could stop a jet landing on an aircraft carrier. But since spiders are hard to farm (they tend to eat each other), it is not possible to collect enough of this silk for commercial use. "Through the marvels of genetic engineering, a company in Canada is now producing the dragline protein in goat's milk!"

Which animal has the strongest bite of all time?

Tops of those measured could be the crocodilians. A 13-foot American alligator has a bite force of about 3000 lbs., says Florida State University paleobiologist Gregory M. Erickson. "Given that there were once 35- to 40-foot crocodilians weighing up to 15,000 lbs., one can imagine them the greatest biters ever."

Other contenders might be the 45- to 60-foot-long extinct Megalodon, a bigger relative of the modern great white shark, with seven-inch teeth compared to a Great White's two inches, says University of Tampa evolutionary biologist Mason Meers. The Megalodon's triangular and serrated teeth would sink deep into its prey before it began thrashing its head from side to side. The teeth would have been like a battery of steak knives with a few thousand pounds of force behind them. Florida's lion-sized saber-toothed cat, one of the most efficient predators on record, had four-inch serrated canines plus razor-sharp upper and lower cheek teeth that worked together like a pair of scissors.

But the all-time king is likely Tyrannosaurus Rex, with over 50 teeth, some half a foot long and an inch thick, says Meers. "My research has shown that T. Rex could generate a bite force roughly 53,000 lbs! For perspective, large fire trucks and tanks weigh about 53,000 lbs. Its bite was one of crushing rather than penetrating or stabbing. So when a T. Rex killed a Triceratops, it was probably by crushing neck bones or by ripping out entire mouthfuls of flesh at a time."

What's the best way to outrun a crocodile?

It's been said that your best bet in fleeing a pursuing crocodile on land is to outwit it by running away in a zig-zag fashion. But that's an old myth and would only up your odds of becoming lunch, says British zoologist Adam Britton. The fastest way of putting distance between yourself and a crocodile is a straight-line run. If you're fit, you should easily win the race.

Some wild speed figures, such as 40 mph, are in TV documentaries, but 8 to 10 mph is more like it for a short run for most crocodiles. They can possibly hit 12 to 20 mph in a brief burst of several feet as they launch themselves out of the water. The fact is, crocs are not pursuers but ambushers. To get away, you have to see one coming. "Never underestimate the attacking speed of a crocodile from a standing start."

How high can bugs fly?

Many insects—and other "bugs" like spiders and mites—ride thermals to high altitudes much like soaring vultures, says Virginia Tech researcher Curt Laub. Migrating monarch butterflies have been seen at 4,000 feet. In Africa and the Middle East, migratory locusts fly in swarms millions strong at a mile high.

USDA researcher P. A. Glick flies planes with insect-census traps on the wings. He estimates that there are millions of bug "aeroplankton" above any given square mile of land. Most bugs he caught under 3,000 feet, but one spider unfurled a strand of silk and rode it balloon-like to 15,000 feet.

Thus airborne, says Iowa State entomologist Robert E. Lewis, spiders have been known to ferry on the jet stream across the Atlantic Ocean.

Essentially floating on the wind, many insects ride weather fronts to cross high mountain passes and oceans to different continents. Moving en masse, moths, grasshoppers, and aphids pose a minor hazard to small planes, says University of Arkansas's Timothy Kring. But commercial jets fly much higher, so their problem is more on takeoffs and landings—forcing heavy windshield scrubdowns.

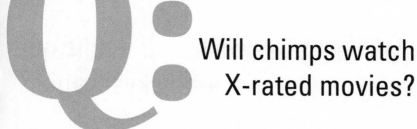

Will chimps watch X-rated movies?

Let's keep this short. Yes, if they are raised by humans and come to think they're human too, said Carl Sagan and Ann Druyan. Raised in the wild, chimps will show no interest in such depictions of Homo sapiens. Now, if the films starred other chimps....

Will a mouse be hurt if it falls out of a skyscraper?

" "Terminal velocity" is the key here.

A skydiver jumping out of a plane accelerates to around 160 mph, depending on body weight and positioning. But he or she then hits a maximum speed, due to air resistance. Small animals have more surface area for their weight, so they generate more air resistance and peak at a much slower speed. Their bodies act as built-in parachutes.

A mouse can fall several thousand feet onto a hard surface and suffer little more than a daze, points out biologist J. B. S. Haldane in his essay "On Being the Right Size." A rat can fall out of an 11th-story skyscraper window and go on about its business. A much longer drop than that would probably do in the rat, but creatures smaller than mice can plunge from very great heights and go virtually unfazed.

REFERENCES

Part 1: *The Body*

Can people grow horns?
Bondeson, Jan. The Two-Headed Boy, and Other Medical Marvels. *Ithaca, NY: Cornell University Press, 2004.*

Can people commit crimes in their sleep?
Shapiro, Colin and Alexander McCall Smith, eds. Forensic Aspects of Sleep. *Chichester, NY: Wiley, 1997.*

Can dreams foretell the future?
Lavie, Peretz. The Enchanted World of Sleep. Trans. Anthony Berris. *New Haven, CT: Yale University Press, 1996.*

Do Siamese twins die together?
Cassill, Kay. Twins: Nature's Amazing Mystery. *New York: Atheneum, 1982.*

Is it possible to will warts away?
Thomas, Lewis. The Medusa and the Snail: More Notes of a Biology Watcher. *New York: Viking Press, 1979.*

Why do faces look they way they do?
Givens, David. Love Signals: How to Attract a Mate. *New York: Crown Publishers, 1983.*

Are there people that don't have "flatulent moments"?
Bolin, Terry and Rosemary Stanton. Wind Breaks: Coming to Terms With Flatulence. *New York: Bantam Books, 1995.*
Paulos, John Allen. Innumeracy: Mathematical Illiteracy and Its Consequences. *New York: Vintage, 1990.*

Can dreams help you stop smoking or lose weight?
Baron, Robert A. Psychology. *2nd Edition. Boston: Allyn and Bacon, 1992.*

Can a guy get pregnant?
Bodanis, David. The Body Book: A Fantastic Voyage to the World Within. *Boston: Little Brown, 1984.*

How sensitive are the five senses?
Morris, Charles G. Psychology: An Introduction. 9th Edition. Upper Saddle River, NJ: Prentice Hall, 1996.

Why do you see bright lights when you press on your closed eyelids?
Walker, Jearl. The Flying Circus of Physics. New York: Wiley, 1975.

Are sleepwalkers acting out their dreams?
Lavie, Peretz. The Enchanted World of Sleep. Trans. Anthony Berris. New Haven, CT: Yale University Press, 1996.

Will it ever be possible to freeze people and bring them back to life in the future?
Iserson, Kenneth V. Death to Dust: What Happens to Dead Bodies. Tucson, AZ: Galen Press, 1994.

Why are sumo wrestlers so fat?
Hall, Mina. The Big Book of Sumo: History, Practice, Ritual, Fight. Berkeley: Stone Bridge Press, 1997.

Can a body live without a head?
Dixon, Patrick. The Genetic Revolution. Eastbourne, UK: Kingsway Communications, 1993.

If you're starving, can you eat your clothes?
Barnett, Clive. Doctor in the Kitchen: A Feast of Protective Foods. Blackheath, NSW, Australia: Verand Books, 1998.

Are their "false pregnancies" in men?
Barrett, Deirdre. The Pregnant Man: Cases from a Hypnotherapist's Couch. New York: Times Books, 1998.

Can human stomach acid dissolve an automobile?
Asimov, Isaac. Book of Facts. New York: Bell, 1979.

Why are there equal numbers of men and women?
Gould, Stephen Jay. The Panda's Thumb: More Reflections on Natural History. New York: W.W. Norton, 1980.

Can you drink your own urine if you are dying of thirst?
Argonne (Illinois) National Laboratory. "Newton's Ask a Scientist Service." http://newton.dep.anl.gov/aasquesv.htm.

Part 2: *Love*

Why is kissing so popular?
Tiefer, Leonore. Sex Is Not a Natural Act and Other Essays. *Boulder, CO: Westview Press, 1995.*

Do opposites attract?
Buss, David M. The Evolution of Desire: Strategies of Human Mating. *New York: Basic Books, 1994.*

Why do "the girls get prettier at closing time"?
Baron, Robert A. and Donn Byrne. Social Psychology. *8th Edition. Boston: Allyn and Bacon, 1997.*

How choosy are most people when it comes to dating?
Hatfield, Elaine and G. William Walster. A New Look At Love. *Lanham, MD: University Press of America, 1985.*

How can you tell if a marriage will last?
Myers, David G. Social Psychology. *6th Edition. Boston: McGraw-Hill, 1999.*

Do people flirt the same way in all cultures?
Ackerman, Diane. A Natural History of Love. *New York: Random House, 1994.*

Can a dream lead to true love?
Alvarez, Alfred. Night: Night Life, Night Language, Sleep, and Dreams. *New York: W.W. Norton, 1995.*

Why is love so fickle?
Hatfield, Elaine and G. William Walster. A New Look at Love. *Lanham, MD: University Press of America, 1985.*

Does music contribute to romance?
Miles, Elizabeth. Tune Your Brain: Using Music to Manage Your Mind, Body, and Mood. *New York: Berkley Books, 1997.*

Is it better to dump your lover, or to get dumped?
Aronson, Eliot, Tim Wilson, and Robin Akert. Social Psychology. *5th Edition. Upper Saddle River, NJ: Prentice Hall, 2005.*

Does playing "hard to get" work?
Liebowitz, Michael R. The Chemistry of Love. *Boston: Little, Brown, 1983.*

Do men dream more about their wives, other women, or other men?
Koulack, David. To Catch a Dream: Explorations of Dreaming. *Albany: State University of New York Press, 1991.*

Why is it so hard to talk to someone you're attracted to?
Givens, David. Love Signals: How to Attract a Mate. *New York: Crown Publishers, 1983.*

What was romance like when Shakespeare wrote *Romeo and Juliet*?

Hatfield, Elaine and Richard L. Rapson. Love and Sex: Cross-Cultural Perspectives. *Boston: Allyn and Bacon, 1996.*

Part 3: *Death*

How long can you survive buried alive?

Bondeson, Jan. Buried Alive: The Terrifying History of Our Most Primal Fear. *New York: W.W. Norton, 2001.*

Could vampires really have existed?

Gomez-Alonso, Juan. "Rabies: A Possible Explanation for the Vampire Legend." *Neurology 51, no. 3 (1998): 856–859.*

Can dead bodies literally spin in their graves?

Barber, Paul. Vampires, Burial, and Death: Folklore and Reality. *New Haven, CT: Yale University Press, 1988.*

Can you die from having sex in a dream?

LaBerge, Stephen. Lucid Dreaming. *Los Angeles: J. P. Tarcher, 1985.*

Will AC or DC electricity kill you faster?

Livingston, James D. Driving Force: The Natural Magic of Magnets. *Cambridge, MA: Harvard University Press, 1996.*

Do patients sometimes wake up in the middle of surgery?

Chudler, Erik H. The Neuroscientist Network. *http://faculty.washington.edu/chudler/neurok.html.*

Can a voodoo curse kill?

Coon, Dennis. Essentials of Psychology. *9th Edition. Belmont, CA: Wadsworth, 2003.*

Can you be in a terrible accident without feeling pain?

Livingstone, David. Missionary Travels and Researches in South Africa. *London: J. Murray, 1857.*

Thomas, Lewis. The Medusa and the Snail: More Notes of a Biology Watcher. *New York: Viking Press, 1979.*

Would a penny tossed off the Empire State Building bore through a human skull?

Bloomfield, Louis A. How Things Work: The Physics of Everyday Life. *New York: Wiley, 1997.*

Would a guillotined head feel itself hit the ground?

McIntyre, Dale. "Last Word." New Scientist 168, no. 2269 (Dec. 16, 2000).

Kershaw, Alister. History of the Guillotine. *London: J. Calder, 1958.*

Is it common to die on the toilet?

Hart-Davis, Adam. Thunder, Flush, and Thomas Crapper: An Encyclopedia. *North Pomfret, VT: Trafalgar Square, 1997.*

Can a person die laughing?

Provine, R. R. "Laughter." American Scientist *84 (1996): 38-45.*

Ramachandran, V. S. and Sandra Blakeslee. Phantoms in the Brain: Probing the Mysteries of the Human Mind. *New York: William Morrow, 1998.*

Can dreams serve as early warning signs of illness?

Van de Castle, Robert L. Our Dreaming Mind. *New York: Ballantine Books, 1994.*

Schneider, Daniel E. Revolution in the Body-Mind. *Vol. 1,* Forewarning Cancer Dreams and the Bioplasma Concept. *East Hampton, NY: Alexa Press, 1976.*

Is it possible to be scared to death?

Lown, Bernard. The Lost Art of Healing. *Boston: Houghton Mifflin, 1996.*

Coon, Dennis. Essentials of Psychology. *9th Edition. Belmont, CA: Wadsworth, 2003.*

Is it unreasonable to be afraid of being buried alive?

Iserson, Kenneth V. Death to Dust: What Happens to Dead Bodies. *Tucson, AZ: Galen Press, 1994.*

Part 4: *Animals*

How is animal sex different from human sex?

Diamond, Jared. Why is Sex Fun? The Evolution of Human Sexuality. *New York: Harper Collins, 1997.*

When lightning strikes a body of water, do all the fish in the water die?

Uman, Martin A. All About Lightning. *New York: Dover, 1986.*

Can a dog tell if its owner is sick?

Coren, Stanley and Janet Walker. What Do Dogs Know? *New York: Free Press, 1997.*

Could a human outrun a dinosaur?

Alexander, R. McNeill. Animals. *New York: Cambridge University Press, 1990.*

Do chimps recognize themselves in mirrors?

Lippa, Richard A. Introduction to Social Psychology. *2nd Edition. Pacific Grove, CA: Brooks/Cole, 1994.*

Do animals like alcohol?

Siegel, Ronald K. Intoxication: Life in Pursuit of Artificial Paradise. *New York: Pocket Books, 1989.*

Why do dogs make such great pets?

Murchie, Guy. The Seven Mysteries of Life: An Exploration in Science and Philosophy. *Boston: Houghton Mifflin, 1978.*

Clutton-Brock, Juliet. A Natural History of Domesticated Mammals. *2nd Edition. New York: Cambridge University Press, 1999.*

Coren, Stanley. The Intelligence of Dogs: Canine Consciousness and Capabilities. *New York: Free Press, 1994.*

Can you give sick dogs and cats human medicine?

Dworetzky, John. Psychology. *6th Edition. Pacific Grove, CA: Brooks/Cole, 1997.*

Are there "stealth" bugs?

Hughes, Howard C. Sensory Exotica: A World Beyond Human Experience. *Cambridge, MA: MIT Press, 1999.*

Why is there more superstition about cats than dogs?

Clutton-Brock, Juliet. A Natural History of Domesticated Mammals. *2nd Edition. New York: Cambridge University Press, 1999.*

How do dolphins see and then tell their friends about the sights beheld?

Hewitt, Paul. Conceptual Physics. *9th Edition. San Francisco: Addison-Wesley, 2005.*

What's the best way to outrun a crocodile?

Britton, Adam. "Crocodilians: Natural History and Conservation." *http://www.crocodilian.com.*

Will chimps watch X-rated movies?

Sagan, Carl and Ann Druyan. Shadows of Forgotten Ancestors: A Search for Who We Are. *New York: Random House, 1992.*

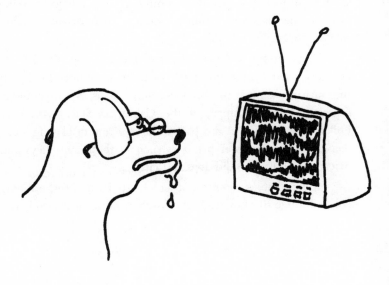

Q: Do chimps recognize themselves in mirrors?

A t first they think they're seeing another chimp and will try to make contact. But within days, they'll begin making faces in the glass and trying to glimpse parts of their bodies they've never seen before.

When chimps were anesthetized and had one ear and one eyebrow dyed bright red, many of them later noticed this in the mirror, fussing at the strange-colored areas, says psychology professor Richard Lippa, of the California State University, Fullerton. Orangutans also passed the dye test, displaying "self-recognition," but lower primates such as rhesus monkeys and macaques flunked, as did dogs and rats.

Human babies catch on at age 15 to 18 months, noticing colored rouge sneakily daubed onto their noses by researchers.